Redistribution Reactions

Redistribution Reactions

J. C. LOCKHART

Department of Inorganic Chemistry
University of Newcastle
Newcastle upon Tyne, England

 1970

ACADEMIC PRESS   New York and London

ACADEMIC PRESS, INC.
111 Fifth Avenue, New York, New York 10003

United Kingdom Edition published by
ACADEMIC PRESS, INC. (LONDON) LTD.
Berkeley Square House, London W1X 6BA

LIBRARY OF CONGRESS CATALOG CARD NUMBER: 75-127692

PRINTED IN THE UNITED STATES OF AMERICA

Contents

PART I

PART II

Preface

There has been a growing recognition of the prevalence of redistribution reactions in inorganic chemistry. Where once they were discovered by chance, they are now systematically investigated, particularly in main group chemistry. Their peculiar fascination seems to lie in kinetic and thermodynamic characteristics, particularly of the random scrambling variety of reaction. Considerable information has been published concerning the kinetics and thermodynamics of redistribution and studies have extended to at least sixty of the elements of the Periodic Table. The need arose for a collective treatment of known data to act as a compendium and to stimulate further research, especially for the less well-documented d and f block elements. This book is intended to fill the existing need. It will provide a convenient reference for research personnel already working on redistribution, and for a more general range of research workers in inorganic chemistry who require information on scrambling for a specific element. It will be of considerable use to those working on inorganic polymer systems, the design of which can be greatly assisted by application of the concepts of structural reorganization. It is hoped that all readers interested in practical inorganic chemistry will become aware of the possibilities of scrambling inherent in their own personal research.

Experimental methods for the study of redistribution phenomena are discussed in the first part, and results are discussed systematically under Periodic Group headings in the second part. The book is selective, emphasizing those reactions for which quantitative information as to kinetics or thermodynamics is available, to the exclusion of many observations which were purely qualitative in nature. Valuable structural information (vibrational frequencies, chemical shifts, etc.) is often included where this is quantitative.

This is the first book to give a comprehensive coverage of redistribution phenomena for the entire Periodic Table. Literature surveys were conducted up to 1969 as far as possible. Reaction energies, activation energies, entropies, etc. have all been expressed in SI units. Temperature values are in degrees Celcius, unless otherwise noted.

The author is grateful to many colleagues for advice and assistance in the preparation of this book, and in particular to Dr. J. R. Blackborow who read most of the manuscript.

Conversion Table for SI Units*

CGS	SI
Energy, Work	
1 calorie (thermochemical)	4.184 joules (exactly)
1 electron volt	1.602×10^{-19} joule
1 entropy unit	4.184 joules per degree Kelvin
Length	
1 angstrom	0.1 nanometer (exactly)
1 micron	1 micrometer (exactly)
Pressure	
1 conventional millimeter of mercury	133.32 newtons per square meter (exactly)
Quantities of Electricity and Magnetism	
1 gauss	10^{-4} tesla (exactly)

* Only units relevant to the text of this volume have been included.

Redistribution Reactions

PART I

1

Introduction

Redistribution reactions were so named by G. Calingaert and H. A. Beatty of Ethyl Gasoline Co., who pioneered their study. In 1939 they discovered a typical example of a slow, random reaction and realized its generality [1]. All the examples which Calingaert and co-workers investigated were perforce of the slow type, no rapid reaction techniques being available. Today many hundreds of examples of the redistribution reaction are known and, thanks to the enormous battery of physical techniques available to modern chemistry, we can detect redistributions in ways undreamed of in 1939. It is not necessary now for their recognition that they be slow and random. In this book the present state of research on redistributions is surveyed. Application of experimental techniques to the study of redistribution is discussed critically in the first section, which is intended as a practical guide for research workers. In the second section, a survey of known reactions as they occur throughout the periodic table is made, and the final chapter assesses the interpretation of reactions, their thermodynamic and kinetic implications. The terms scrambling, redistribution, and exchange are used interchangeably.

The best working definition [2] of a redistribution is that of H. Skinner: "A redistribution reaction is one in which bonds change in relative position but not in type." We will employ an empirical classification of redistribution reactions for the purposes of this book, quoting specific rather than general

1

examples, which are intended to supplement Skinner's definition. Redistribution of two or more different ligands on one central atom is the first major category (e.g., equilibrium 1.1). Almost any metal or metalloid may

$$B(OBu)_3 + B(OEt)_3 \rightleftharpoons B(OBu)_2(OEt) + B(OBu)(OEt)_2 \qquad 1.1$$

replace boron, the valence of the central atom may be more or less than three, and the alkoxy groups may be replaced by any suitable monofunctional ligand. Most of the quantitative kinetic and thermodynamic data concerning redistribution reactions has been assembled for this category, and it is a main concern of this book. A second and wider category is of two or more different ligands scrambling on two or more different central atoms, exemplified by equilibrium 1.2, a typical Swart's reaction. Few examples

$$AsF_3 + PCl_3 \rightleftharpoons AsCl_3 + PF_3 \qquad 1.2$$

have been studied in which more than two different ligands or central atoms are involved in scrambling, and clearly these are increasingly difficult to interpret because of the multiplicity of products. In a third major category where the central atoms and at least one of the ligands are polyfunctional, and the ligand may be simultaneously bonded to more than one individual central atom in a stable fashion, the possibility of polymer formation arises. Difunctional ligand with difunctional central atoms can give rise to ring and chain polymers, as, for example, in equilibrium 1.3, where $Me_2Ge{=}$ and $-S-$ are difunctional [3]. Polymerization does not, however, always occur

$$\begin{array}{cc} [Me_2GeS]_3 + Me_2GeCl_2 \rightleftharpoons Me_2Ge \cdot Cl(SGeMe_2)_nCl, & n = 1\text{--}6 \\ \text{(ring)} & \text{(chain)} \end{array} \qquad 1.3$$

with polyfunctional reagents. The reaction in which β-diketonate ligands, e.g., acetylacetonate (acac) and trifluoroacetylacetonate (tfac), scramble on zirconium, for instance [4, 5], is really of the first category (Eq. 9.4). Trifunctional central atoms can give nets, and central atoms of higher functionality can give three-dimensional polymers.

$$Zr(acac)_4 + Zr(tfac)_4 \rightleftharpoons Zr(acac)_3(tfac) + Zr(acac)(tfac)_3, \qquad \text{etc.} \qquad 1.4$$

We have now clothed Skinner's definition in more familiar chemical terms, but it has further significance. If bonds do not alter in type, the corresponding bond energies should not alter appreciably, that is no change in enthalpy may be expected—a thermoneutral reaction in fact. For a thermoneutral reaction, free energy will depend only on the entropy term, and a purely random distribution of ligands will result. The thermodynamic significance of this is patent—nonrandom reactions are not thermoneutral and the nonzero enthalpies reflect changes in bond strengths. Out of studies of sizable redistribution enthalpies have come new predictive bond energy equations [6]. It was the approximately random distribution of alkyl ligands in tetraalkyllead mixtures which prompted Calingaert's exploration of the redistribution reaction [1]. Initially reactions with considerable half-lives (hours, days, or more) were found to give random distribution. Such reactions are simplest to follow, since the mixed products can usually be isolated and analyzed separately, and both kinetic and thermodynamic

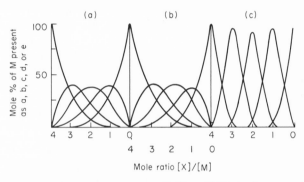

Fig. 1. Equilibrium composition in mixtures of MX_4 and MY_4 for the range of mole ratios [X]/[M] = 0–4. See Eq. 1.5. (a) Statistical distribution. (b) Real distribution M = Ni, X = PF_3, Y = CO [7]. (c) Real distribution M = Si, X = NMe_2, Y = Cl [8].

data are readily accessible. Recently, it has become possible to study by physical techniques, reactions with half-lives of the order of minutes or even less even when these are random, despite the fact that the mixed compounds would not normally be isolated as discrete compounds from such a labile equilibrium. Such fast random reactions are very instructive examples for teaching the importance of kinetics and thermodynamics in the separation and control of reaction products. The very slow random reactions at equilibrium serve to demonstrate thermodynamic control, since little reversal can occur during chemical separation.

In Fig. 1, three composition diagrams are given for reaction 1.5 for the range of mole ratios [X]/[M] from 0 to 4, showing the amounts of each

$$2MX_4 + 2MY_4 \rightleftharpoons MX_3Y + 2MX_2Y_2 + MXY_3$$
$$a e b c d \qquad 1.5$$

compound a, b, c, d, and e present at equilibrium for the specified ratio of [X]/[M]. In Fig. 1a, the statistical distributions of the five species are given for the random case. Fig. 1b shows the near-random distribution observed in the reaction of $Ni(CO)_4$ with $Ni(PF_3)_4$ [7] and Fig. 1c shows the exothermic reaction of $Si(NMe_2)_4$ and $SiCl_4$ with a distribution of a–e very far from random [8].

The composition to be expected at equilibrium may be calculated statistically for a mixture of compounds with central atom M (functionality n) and m ligands X, Y, Z, etc. The number of different compounds which can form is

$$(n + m - 1)!/n!(m - 1)!$$

Let the mole fraction of X groups be f_X, of Y groups be f_Y etc.; then the concentration of the mixed compounds $MX_aY_bZ_c \cdots$ etc. (where $a + b + c + \cdots + j = n$) will be

$$[n!(f_X)^a(f_Y)^b(f_Z)^c \cdots (f_Q)^j]/a!b!c! \cdots j!$$

This is discussed by Calingaert and Beatty [1] and Moedritzer [9]. For two different central atoms, Moedritzer compares experimental findings with computed data for the random case in diagram form [9].

Although the thermoneutral example in Fig. 1b is that of a transition metal [7], and the exothermic example in Fig. 1c is that of a main group element [8], the positions might well have been reversed. Enthalpies from large positive through large negative are possible for most multivalent elements (as central atoms), depending rather on the substituents than on the element's position in the periodic table. There has been more quantitative work on scrambling on main group central atoms, thanks to the work principally of Van Wazer and Skinner and their respective schools. This has generally been on nonaqueous systems, and has resulted in a fairly broad picture, mainly of the thermodynamics of scrambling on main group elements. The kinetics seem to depend much more than the thermodynamics on the nature of the central atom(s), as well as that of the ligands. Four-center transition states are thought to be prevalent in main group chemistry. Much of the work on transition metals has been done for aqueous solutions and the mechanisms are predominantly ionic or dissociative in nature. Recent work on carbonyls [7] shows that there is little fundamental difference between main group and transition metal redistributions of covalent molecules.

Redistribution reactions were the subject of a recent conference, the proceedings of which have now been published [10].

REFERENCES

1. G. Calingaert and H. A. Beatty, *in* "Organic Chemistry, An Advanced Treatise," Vol. II, p. 1806. Wiley, New York, 1950.
2. H. A. Skinner, *Rec. Trav. Chim.* **73**, 991 (1954).
3. K. Moedritzer and J. R. Van Wazer, *J. Amer. Chem. Soc.* **87**, 2300 (1965).
4. A. C. Adams and E. M. Larsen, *Inorg. Chem.* **5**, 228 (1966).
5. T. J. Pinnavia and R. C. Fay, *Inorg. Chem.* **5**, 233 (1966).
6. G. A. Nash, H. A. Skinner, and W. F. Stack, *Trans. Faraday Soc.* **61**, 640, 2122 (1965).
7. R. J. Clark and E. O. Brimm, *Inorg. Chem.* **4**, 651 (1965).
8. K. Moedritzer and J. R. Van Wazer, *Inorg. Chem.* **3**, 268 (1964).
9. K. Moedritzer, *Advan. Organometal. Chem.* **6**, 171 (1968).
10. Redistribution reactions in chemistry, *Ann. N.Y. Acad. Sci.* **159**, 1 (1969).

2

Experimental Techniques

Any experimental technique which is capable of detecting all compounds in a redistribution reaction may in principle be used, from simple titration to the most sophisticated physical method. Usually physical techniques are directed at the reaction mixture itself, but separation of components is also employed. A series of techniques, electronic, vibrational and nuclear magnetic resonance spectroscopy, X-ray crystallography, mass spectroscopy, polarography, chromatography, the phase diagram and the use of labels (chemical and radioactive) is covered in Chapter 2, from two angles— the first a discussion of the application of the technique to redistribution in particular, and the second a demonstration of the technique for a concrete example.

I. Spectroscopy

A. *Light Absorption Spectra*

Absorption spectra are caused by the absorption of incident radiation by a substrate; transitions between energy levels in the substrate result. In the ultraviolet and visible spectral region [50,000–12,500 cm^{-1} (15×10^{14}–37.5×10^{13} Hz) frequency range] transitions are between electronic energy levels and in the infrared [12,500–20 cm^{-1} (37.5×10^{13}–60×10^{10} Hz) frequency range] between vibrational and rotational levels. Raman spectra are somewhat different in that they are produced by loss or gain of quanta of vibrational or rotational energy by a substance when a beam of mono-chromatic light is passed through it. A small portion of the light is scattered with change in frequency (the new frequencies are the Stokes and anti-Stokes lines) corresponding to the vibrational or rotational energy trans-itions caused. The selection rules are somewhat different from those of infrared spectroscopy, but the frequency region covered is similar and the two are often complementary. Frequencies absorbed are characteristic of the substrate and in one or other region of the spectrum may lead to identi-fication of species and often to their quantitative analysis. If the light absorption concerned follows Beer's law,* then the intensity of absorption at a particular frequency for a particular compound may be used as a meas-ure of its concentration. Instrumentation is better developed for the quanti-tative measure of concentration in ultraviolet and visible spectroscopy than in infrared; quantitative work is also possible with recording Raman instru-ments. At least one region of the spectrum can usually be found in which it is possible to distinguish the components of a redistribution reaction and to analyze separately for the concentration of each. A time factor can be important in the spectroscopic study of a reaction mixture, such as that in redistribution. If an absorbing substrate is involved in any isomerization or chemical reaction for which its lifetime (τ) for conversion between the two chemical forms is of the order of, or much shorter than, the inverse of the difference between corresponding absorption frequencies (expressed as hertz or reciprocal seconds) of the two chemical forms, then the observed spectrum will exhibit just one such frequency at the (weighted) mean of the two expected.

After this general discussion of the origin of spectra, we now consider vibrational and electronic spectra in more detail, noting how they may best be applied to redistribution reactions.

1. VIBRATIONAL SPECTROSCOPY

a. General detection of redistribution

The vibrational spectrum of an equilibrium mixture is often of great help in the general detection of redistribution reactions, particularly where the

* The Beer–Lambert law states that the log of the ratio of intensity of emergent (I_0) to incident (I) light (optical density, D) is proportional both to the concentration (c) and thickness (d) of the light-absorbing substance. $D = \log_{10}(I_0/I) = \epsilon c d$. The proportionality constant, ϵ, is known as the extinction coefficient of the material in question.

equilibrium is labile. The symmetry of the end components MX_n and MY_n in reaction 2.1 will be the same, but usually higher than that of the mixed

$$MX_n + MY_n \rightarrow MX_{n-1}Y + \cdots + MXY_{n-1} \qquad\qquad 2.1$$

compounds. Accordingly the selection rules will differ for end and mixed components and most usually a greater number of fundamental frequencies will be allowed in the spectra of the less symmetrical compounds. When the equilibrium is labile and only the symmetrical components can be isolated pure, their spectra can be determined separately. If then the spectrum of their mixture is taken, it will contain, apart from the superposed spectra of the two end components, additional absorption bands, corresponding to the extra frequencies allowed for the less symmetrical components. Sometimes the mixed and end components have the same number of allowed frequencies, but even here the simple mass effect (combined with the electronic effect) of the substituents often differentiates between the end and mixed components, since the location of frequencies associated with motion of substituents involved in scrambling, will vary according to the mass and bonding strength of the substituent. Cases do arise however in which the skeletal symmetries of the reagents and products are the same and the mass effect is unimportant: for these, vibrational spectroscopy would be useless to detect scrambling. The symmetry class of observed frequencies in a redistribution mixture is very helpful in making the necessary assignments. For these purposes both infrared and Raman spectroscopy are used.

Historically, Raman spectroscopy was the favored and in fact the first technique by which labile redistribution reactions were observed. Where the reaction concerned redistribution of heavy substituents (e.g., Cl, Br, I) on heavy central atoms, the metal–halogen frequencies were accessible to Raman spectroscopy, but not to infrared spectrometers of the limited range then generally available. Infrared spectrometers now commercially available operate to frequencies as low as 20 cm^{-1} (60 × 10^{10} Hz) and the infrared technique is often simpler to practice. The analysis of the vibrational spectrum of the reaction mixture is facilitated by the complementary use of both techniques, which is the ideal approach.

Delwaulle investigated scrambling of halogen on germanium, silicon, and tin using Raman spectroscopy. The tetrahalides of Group IVB are of tetrahedral (T_d) symmetry, and selection rules for Raman spectra permit four fundamental bands in the spectrum (one polarized). In mixtures of $SiCl_4$ and SiI_4, scrambling to give $SiICl_3$, SiI_2Cl_2, and $SiClI_3$ is expected. These should have respectively C_{3v}, C_{2v}, and C_{3v} symmetry, with 6, 9, and

6 fundamental Raman absorptions allowed. Figure 2.1 contains the correlation chart for Raman frequencies observed by Delwaulle for silicon chloride iodides [1].

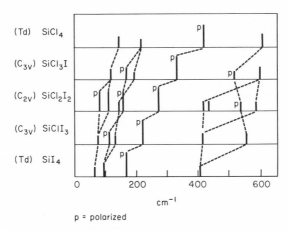

Fig. 2.1. Correlation of Raman frequencies for the set of mixed $SiCl_{4-x}I_x$, where $x = 0–4$ [1].

Obviously, the simpler the vibrational spectra of the reagents, the more suitable these techniques are for detection of redistribution. A complicated spectrum is obtained for substituents such as dialkylamino [2, 3] or alkoxy [4]. Where these scramble with substituents of another type (e.g., Eq. 2.2)

$$(Me_2N)_3B + BCl_3 \rightleftharpoons Me_2NBCl_2, \qquad \text{etc.} \qquad 2.2$$

the spectrum of the mixed compounds would be distinctive in regions other than those of the alkylamino group frequencies. When two different alkoxy groups scramble on boron however, no change in skeletal symmetry is expected, and the mass effect is insignificant on the B–O part of the spectrum so that mixed esters still effectively have D_{3h} skeletal symmetry at boron [4].

The mere observation of new frequencies in mixtures in the anticipated range is good evidence for the occurrence of a redistribution reaction. Other information may be gleaned from a more extensive investigation of the spectrum.

b. Force constants

In a complete analysis of the spectrum, including assignment of each frequency to a particular vibrational mode, the force constants for the stretching of the bonds involved in the reaction scheme (MX and MY) will be obtained for each compound. For the components of a thermoneutral reaction, the bond strength MX (and hence the stretching force constant) should be constant throughout the series MX_n, MX_{n-1}, MX_{n-2}, etc. The extent to which force constant varies is then a measure of the departure from randomness in this reaction. This approach is crude since, of course, the force constant measures the resistance of a bond to stretching (Hooke's law) and this is not quite the same thing as the resistance to bond fission, which is the thermodynamic bond strength [5]. The information should nevertheless be valuable, especially if it can be compared with true thermodynamic bond strengths.

c. Thermodynamic properties

The complete assignment of the fundamental modes in a vibrational spectrum also enables the calculation of various thermodynamic quantities for the compound: they are heat capacity $C_p{}°$; heat content function $(H° - H_0°)/T$; entropy $S°$; and free energy function $(G° - H_0°/T)$; they are calculated by methods based on statistical thermodynamics, which are extensively treated in text books and literature [6–10]. For the most general case of an asymmetric top (harmonic oscillator and rigid rotator) the separate translational, rotational, and vibrational contributions to each thermodynamic function may be calculated from Eqs. 2.3–2.15 derived by the methods of statistical thermodynamics. The molecular parameters required are the molecular mass M, moments of inertia I_A, I_B, I_C, symmetry number σ, and fundamental vibrational frequencies ν (cm^{-1}); the functions are obtained to a close approximation by summation of the separate contributions due to vibration, rotation, and translation:

Heat capacity:

$$C_p = C_{p\,(\text{trans})} + C_{p\,(\text{rot})} + C_{p\,(\text{vib})} \qquad\qquad 2.3$$

$$C_{p\,(\text{trans})} = 4.9680 \qquad\qquad 2.4$$

$$C_{p\,(\text{rot})} = 2.9808 \qquad\qquad 2.5$$

$$C_{p\,(\text{vib})} = 1.9872 \sum_j [x_j{}^2\, e^{x_j}/(e^{x_j} - 1)^2] \qquad\qquad 2.6^*$$

Heat content function:

$$(H° - H_0°)/T = [(H° - H_0°)/T]_{\text{trans}} + [(H° - H_0°)/T]_{\text{rot}}$$
$$+ [(H° - H_0)°/T]_{\text{vib}} \qquad\qquad 2.7$$

* $x_j = \nu_j hc/kT$, where ν is the jth fundamental vibrational frequency (cm^{-1}); h and k are the Planck and Boltzmann constants.

$$[(H° − H_0°)/T]_{trans} = 4.9680 \tag{2.8}$$

$$[(H° − H_0°)/T]_{rot} = 2.9808 \tag{2.9}$$

$$[(H° − H_0°)/T]_{vib} = 1.9872 \sum_j [x_j/(e^{x_j} − 1)] \tag{2.10}$$

Entropy:

$$S° = S°_{trans} + S°_{rot} + S°_{vib} = −(G° − H°)/T \tag{2.11}$$

Free energy function:

$$−(G° − H_0°)/T = −(G° − H_0°)/T_{trans} − (G° − G_0°)/T_{rot} \\ − (G° − H_0°)/T_{vib} \tag{2.12}$$

$$−[(G° − H_0°)/T]_{trans} = 6.8635 \log M + 11.4392 \log T − 7.282 \tag{2.13}$$

$$−[(G° − H_0°)/T]_{rot} = 2.2878 \log(I_A I_B I_C × 10^{117}) + 6.8635 \log T \\ − 4.5757 \log σ − 3.0140 \tag{2.14}$$

$$−[(G° − H_0°)/T]_{vib} = 1.9872 \sum_j \ln(1 − e^{−x_j}) \tag{2.15}$$

Tables of vibrational contributions to thermodynamic functions, usually given as ratios, e.g., S/R for particular values of x_j, are available [8, 9]. Somewhat different equations are required for spherical symmetrical tops

TABLE 2.1

$−Δ(G° − H°)/T$ for Reactions 2.16 and 2.17, in an Ideal Gas State, at 1 atm Pressure[a] (101.325 kN m^{-2})

°K:	300	500	700	800	1000	1300	
	15.4	14.9	13.1	15.2	14.4	15.0	(Eq. 2.16)
	14.0	13.9	13.9	13.8	14.4	13.8	(Eq. 2.17)

[a] Units, J °K^{-1}.

and linear molecules. Equations 2.3–2.15 give thermodynamic data in cal mole^{-1}, which can be converted to SI units by the relation 1 cal (thermochemical) = 4.184 J. For instance, the function $Δ(G° − H_0°)/T$ for reactions 2.16 and 2.17

$$B^{11}Br_3 + B^{11}Cl_3 \rightleftharpoons B^{11}Br_2Cl + B^{11}BrCl_2 \tag{2.16}$$

$$B^{11}F_3 + B^{11}Cl_3 \rightleftharpoons B^{11}F_2Cl + B^{11}FCl_2 \tag{2.17}$$

has been obtained from literature values of thermodynamic quantities calculated by statistical methods for each compound [11, 12]. There is no significant variation with temperature (Table 2.1). The general effect of temperature on each of the redistribution reactions 2.16 and 2.17, will be to increase the concentration of mixed halide, the entropically favored product. A knowledge of the enthalpy of reaction at any point of the temperature ranges quoted would enable the prediction of the complete free energy change in these ranges.

d. Rates and equilibria

Semiquantitative measurements of both rates and equilibria for particular reactions have been made with vibrational spectra, but neither the infrared nor the Raman technique lends itself particularly to this type of study. In the first place, instrumental failings often do not permit Beer's law to hold for IR work; second, there are considerable difficulties in thermostatting reaction mixtures at temperatures other than high ones, because the samples have to be rather close to the source of radiation, which of course is radiating considerable heat.

2. ELECTRONIC SPECTRA

When the electronic spectra of various components of a redistribution reaction differ sufficiently, they provide an acceptable physical method of exploring the reaction. The technique is likely to have its greatest application in solution chemistry of transition metal complexes. Here distinctive wavelengths (λ_{max}) of maximum absorption and extinction coefficients (ϵ_{max}) of components are useful in the analysis of mixtures, although in solution the bands may sometimes be broad and overlapping. In principle one can analyze for the concentrations of an n-component mixture provided one knows the ϵ_{max} and λ_{max} of $n - 1$ of the components in isolation and Beer's law applies to each, and in practice computers are often used to perform the analysis. The relation of extinction ϵ_{max} to concentration means that both equilibrium and kinetic data may be determined.

Possible equilibria of metal complexes in solution include the formation of the complexes from metal and ligands as well as possible redistribution reactions between them, e.g.,

$$M + X \rightleftharpoons MX \qquad M + Y \rightleftharpoons MY$$

$$M + 2X \rightleftharpoons MX_2 \qquad M + 2Y \rightleftharpoons MY_2$$

$$MX_2 + MY_2 \rightleftharpoons 2MXY$$

and it is probable that absorption characteristics for all of these species (λ_{max} and ϵ_{max}) will be required. The Job's plot method is commonly employed to determine equilibrium constants. Kinetics of reaction may be followed in the spectrometer cell directly (down to half-lives of a few hundredths of a second) or after quenching the reaction mixture to stop the reaction.

A recent application of UV spectroscopy to the equilibria between mercuric halides in aqueous solutions allowed the equilibrium constants for redistributions 2.18–2.20 to be determined [13]. This system is unusual

$$\log K = 1.14 \pm 0.11 \qquad HgCl_2 + HgBr_2 \rightleftharpoons 2HgClBr \qquad 2.18$$

$$\log K = 1.07 \pm 0.08 \qquad HgBr_2 + HgI_2 \rightleftharpoons 2HgBrI \qquad 2.19$$

$$\log K = 1.35 \pm 0.17 \qquad HgCl_2 + HgI_2 \rightleftharpoons 2HgClI \qquad 2.20$$

$$\log K_{stat} = 0.60.$$

in that the concentrations of metal ion and free halide ions are negligible and the only species of importance are the molecular halides. Each reaction studied thus has only three components to complicate the spectrum.

The ϵ_{max} and λ_{max} of the mercury homohalides can be determined for each in isolation. The spectra of their mixtures are not simply the sum of the absorptions of the two starting halides and the residual absorption is that of the mixed halide. Take reaction 2.18. At very high ratios, $R = [HgCl_2]/[HgBr_2]$, all bromide in the mixture will be present as HgClBr. At very low ratios, R, all chloride will be tied up as molecular HgClBr. One can thus determine ϵ_{max} and λ_{max} for the heterohalide. The parameter x where x is the fraction of Hg present as HgClBr and

$$x = \left[1 - \left(1 - \frac{4R(1 - 4K^{-1})}{(1 + R)^2}\right)^{1/2}\right](1 - 4K^{-1}) \qquad 2.21$$

is plotted against $\log R$ and compared with simulated curves for a series of possible values of K to obtain the best approximation to K.

A full account of the determination of stability constants by this technique is given by Rossotti and Rossotti [14].

B. Nuclear Magnetic Resonance

This is essentially another kind of absorption spectroscopy in which a nucleus of nonzero spin, subject to a strong magnetic field (H) undergoes a net absorption of applied energy at radio frequencies, which produces transitions from lower to higher nuclear energy levels. These energy levels correspond to different alignments of the nuclear spin vector with respect to the applied field H, and the application of the weaker radio frequency oscillating at right angles to the applied field causes transitions between the nuclear spin levels. When the nuclei are present in a chemical environment they are shielded from the full strength of the applied field. The difference between the applied field H_0 and the field experienced at the nucleus is a very delicate function of the chemical environment of the nucleus and can be used to differentiate between chemically different environments. The chemical shift is a convenient parameter used to express this difference and is defined as δ, where

$$\delta = (H_{obs} - H_{ref})/H_{ref} \times 10^6 \qquad\qquad 2.22$$

H_{obs} is the strength of the applied field observed for the environment in question and H_{ref} is the strength of the field at any suitable standard chemical environment of the same nucleus; δ is expressed as parts per million (ppm) of the applied field. Reference standards in common use include tetramethylsilane (TMS) for ^1H nuclei, methyl borate for ^{11}B nuclei, phosphoric acid for ^{31}P nuclei, $CFCl_3$ for ^{19}F nuclei, etc.

Nuclear magnetic resonance is one of the most powerful aids in the study of redistribution reactions. Two environments not detectably different to other artificial eyes such as vibrational spectroscopy may give rise to different chemical shifts in NMR spectra. General detection of redistribution is simple if the molecules taking part in the reaction contain nuclei suitable for NMR techniques, because scrambled products will give signals of different chemical shift from the starting materials.

Nuclear magnetic resonance has been used extensively in thermodynamic and kinetic studies of redistribution, especially by Van Wazer and his co-workers. See, for example, a résumé of their philosophy as it applies to ^{31}P NMR [15]. The NMR time scale is relevant for these purposes. For a thermodynamic study, it is desirable that the reaction be sufficiently slow in order for the chemical shift difference between the exchanging species (Δ is expressed in hertz) to be much greater than $1/\tau$, where τ is the mean exchange lifetime of the nucleus (sec) between different environments, $\Delta \gg 1/\tau$. In this limiting situation, the area of the signal observed is proportional to the concentration of the nucleus causing the signal, and peak areas may be used to determine equilibrium constants. This situation is also suitable for kinetic work in which signal area is used as a measure of concentration and the peak area changes slowly with time. The relation between concentration and time then gives the kinetic form of the reaction.

The especial contribution of NMR to redistribution kinetics, however, is in the rate situation where the chemical shift difference between the exchanging species, Δ, is approximately the same as or very much less than $1/\tau$:

$$\Delta \approx 1/\tau \quad \text{or} \quad \Delta \ll 1/\tau \qquad 2.23$$

When such a time scale applies to the redistribution reaction in question, sharp individual signals for the exchanging chemical environments will no longer be observed. Instead, the nucleus in resonance experiences an "average" environment—the mean of those between which it is exchanging. There is considerable use made of the related and somewhat special rate situation in which Δ, the separation between peaks of a multiplet due to

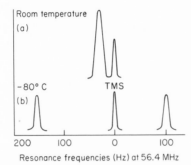

Fig. 2.2. Schematic representation of ^{1}H NMR spectrum of TlMe$_3$ in toluene: (a) $\Delta \ll 1/\tau$; (b) $\Delta \gg 1/\tau$. ^{205}Tl–^{1}H coupling is seen in (b), but in (a) the reaction is fast enough to obliterate it.

spin–spin coupling, is $\ll 1/\tau$. The simple case of a doublet is depicted in Fig. 2.2. The ^{1}H nuclei in the methyl groups are coupled to the thallium nuclei with spin $\frac{1}{2}$, and so appear as a doublet in Fig. 2.2b, where $\Delta \gg 1/\tau$. The useful time scale here is when $\Delta \ll 1/\tau$ and exchange of methyl groups on thallium is fast (Fig. 2.2a). The components of the doublet are now averaged out. The use of this collapse of signals is the same whether the separation is a chemical shift difference or due to spin coupling. The signals will show characteristic shapes, ranging from broadened signals where $\Delta \approx 1/\tau$ to a sharp line at the weighted mean of the expected signals in the limit of very fast exchange, $\Delta \ll 1/\tau$. Values of τ may be extracted from line-shape data [15–17] and can be used to obtain more conventional kinetic parameters, rate constants, and activation energies. Where the time scale of the reaction to be studied is inconveniently fast or slow for the desired measurements, it is in principle possible to change its rate by heating, cooling, catalysts, inhibitors, etc. Many NMR instruments now have standard equipment for temperature variation. Although strict temperature control is achieved, exact temperature measurement is not possible at all times. This uncertainty is probably a major factor in the variability of results from different laboratories using the same system. It is both difficult to read the temperature exactly and to control it to within $\pm 1°$, which is considered a very poor situation for kinetic work. A secondary standard often used is the temperature dependent spectrum of a standard ethylene glycol sample which is measured before and after the kinetic sample in question.

Nuclear magnetic resonance is unsurpassed as a technique simply for demonstrating the existence of redistribution. In the case of the tin tetrahalides, redistribution which for long defied detection was observed by the ^{119}Sn resonance technique of Burke and Lauterbur [18].

The ^{119}Sn magnetic resonance spectra of mixtures of tin halides provide evidence of the redistribution of halogen. This isotope is 8.68 % abundant in natural tin and has a spin of $\frac{1}{2}$ and gives a stronger signal than those of the other magnetic isotopes. At 8.5 MHz, the ^{119}Sn resonances are found at about 5360 G. The solid SnI_4 and the bromide–iodide mixture were dissolved in a few drops of CS_2. These new peaks are found in mixtures of any two tin halides (Cl, Br, I) with approximately the intensities expected for random exchange. In mixtures of all three halides, a further three peaks of appropriate intensity were observed. All the new peaks can be assigned to mixed halides with confidence. The chemical shifts and their assignments are shown in Table 2.2 [18]. Lifetimes of molecules between exchanges must be

TABLE 2.2

^{119}Sn CHEMICAL SHIFTS (ppm) RELATIVE TO THAT OF $(CH_3)_4Sn^a$

$SnCl_4$	150	$SnCl_3I$	551	$SnCl_2BrI$	672
$SnBr_4$	638	$SnCl_2I_2$	951	$SnClBr_2I$	796
SnI_4	1701	$SnClI_3$	1347	$SnClBrI_2$	1068
$SnCl_3Br$	265	$SnBr_3I$	916		
$SnCl_2Br_2$	387	$SnBr_2I_2$	1187		
$SnClBr_3$	509	$SnBrI_3$	1447		

^{a} From Burke and Lauterbur [18].

$>10^{-3}$ sec, and probably $\geqslant 10^{-2}$ sec, but certainly <10 sec. No precautions were taken to exclude moisture and other contaminants. Where the atom-to-ligand bond changes as in this instance, macroscopic shifts are observed.

Lead chloride and lead bromide are isomorphous (orthorhombic, space group V_h with four molecules per unit cell). For a series of mixtures of the two, a discontinuity in the lattice parameters (Fig. 2.3) at 50 mole %

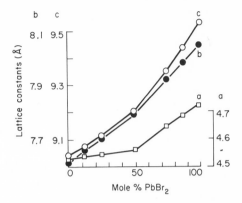

Fig. 2.3. Discontinuities in lattice parameters a, b, and c, at 50 mole % PbBr$_2$ in mixtures of PbCl$_2$ with PbBr$_2$.

composition of each indicates an ordered structure at temperatures well below the melting point and consisting of lead bromide chloride (PbBrCl), which is also isomorphous [23]. However, the phase diagram for the solid–liquid transitions of the system indicates a continuous series of solid solutions (see Fig. 2.4) with no stable mixed compound [23].

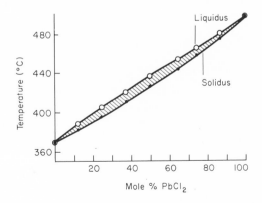

Fig. 2.4. Phase diagram for solid and liquid mixtures of PbCl$_2$ and PbBr$_2$, indicative of solid solutions.

There are two sets of halogen positions in lead chloride (Fig. 2.3) and gradual ordered replacement of the chloride with the greatest spatial freedom occurs up to the 50 mole % bromide composition. Note the more abrupt change in lattice dimensions in Fig. 2.3, occasioned by replacement of the spatially more constrained chloride by the bulky bromide group in the right-hand portion of the figure. Thus, it seems that at room temperature the constraints of the lattice favor an ordered mixed halide structure completely, while at the melting point of the system, the three lead halides ($PbCl_2$, $PbBr_2$, and PbClBr) coexist.

The mixed compound pertungstyl chloride bromide ($WOCl_3Br$) is isostructural with $WOCl_4$ and $WOBr_4$ (body-centered tetragonal, space group $I4$, with two molecules per unit cell) and its lattice parameters are intermediate (Table 2.4) [24]. The structure is thus W—O—W chains, the

TABLE 2.4

ATOMIC COORDINATES AND LATTICE PARAMETERS[a]

	$WOBr_4$	$WOCl_3Br$	$WOCl_4$
U (halogen)	0.261	0.260	0.258
V (halogen)	0.072	0.063	0.067
Z (halogen)	0.090	0.082	0.079
$a_0 = b_0$	8.96	8.52	8.48
c_0	3.93	3.98	3.99

[a] There are two molecules per unit cell.

halogens completing an octahedral coordination around tungsten (structure II). The distribution of halogens on the available sites must in this case be quite random, since no superlattice diffraction spots were observed [24].

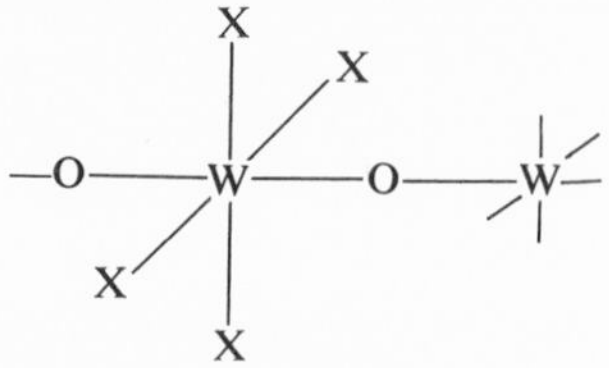

X = Cl or Br

II

III. Mass Spectrometry

The mass spectrum of a volatile redistribution reaction mixture can be used for the unequivocal detection of mixed products, provided that a unique and reproducible mass spectrum appears for each component, and that the cracking patterns of at least the end components are known. Since the mass spectrometer is a very sensitive instrument (resolution of greater than 50,000 is possible on a 10% valley definition with some commercially available machines), quite small amounts of a new scrambled compound may be detected where an unambiguous parent or fragment ion from it is observed.

This might be a signal at a value of m/e where no signal is observed for the end components, or a considerable increase in intensity of some peak already present in the spectrum of an end component relative to that of its parent ion. The appearance of a new metastable peak in a mixture, connecting the decay of a new m/e peak to its daughter fragment is convincing further evidence of scrambling.

Conceivably, redistribution could take place between two parent ions (Eq. 2.25) in the ionization chamber and signals for the resulting mixed

$$MX_n^+ + MY_n^+ \rightleftharpoons MX_{n-1}Y^+ + MXY_{n-1}^+ \qquad 2.25$$

components could be observed. It would be necessary for the reaction to be extremely rapid and since it is an ion–ion collision process, it would be most unlikely except at high pressures.

Mass spectra have also been used to determine thermodynamic data for redistribution reactions in a quantitative fashion. The difficulties attached to this method are that the sensitivity of the instrument to each of the redistributing species (or to all but one of these) must be known, that the intensity of the mass peak for any ion must be proportional to the partial pressure of the compound from which it is derived and that the spectra should be linearly additive. Given these favorable circumstances, an accurate analysis for the concentration of each component of a mixture is possible and hence an accurate determination of K, the equilibrium constant.

Dessy and co-workers designed a simple mass spectra experiment to show that alkyl groups attached to mercury could undergo scrambling. They observed the relevant m/e peaks (Table 2.5) in dimethylmercury, perdeuteriodimethylmercury, and mixtures of the two [25]. This example is somewhat complicated by the number of natural isotopes of mercury

TABLE 2.5

Isotopic Parent Ions Observed in the Mixture of Me_2Hg and
$(CD_3)_2Hg$ after Exchange of Methyl Groups[a]

Compound			m/e values observed			
Me_2Hg	228	229	230	231	232	234
$MeHgCD_3$	231	232	233	234	235	237
$(CD_3)_2Hg$	234	235	236	237	238	240
Hg isotopes	198	199	200	201	202	204

[a] From Dessy *et al.* [25]

present. In a similar study, Fallon and Lockhart [26] have shown the
presence of mixed boric esters from the reaction 2.26. Figure 2.5 shows new

$$B(OEt)_3 + B(OMe)_3 \rightleftharpoons B(OMe)_2(OEt) + B(OEt)_2(OMe) \qquad 2.26$$

peaks observed in the mixture indicative of the presence of mixed esters,
although it is not possible to isolate these mixed compounds on distillation.
New metastable peaks were also observed (Table 2.6) for a typical rearrange-
ment process, involving ejection of olefin, which is commonly observed in

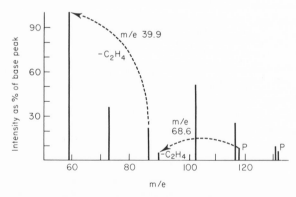

Fig. 2.5. Mass spectra of mixed borates $B(OMe)_2(OEt)$ and $B(OMe)(OEt)_2$, showing
parent ions P and rearrangements.

TABLE 2.6

ORIGIN OF METASTABLE IONS SPECIFIC TO MIXTURES OF ETHYL AND METHYL BORATES
NOT OBSERVED IN EITHER PARENT ALONE[a]

Observed m/e	Calculated m/e	Process
68.6	68.6	$B(OMe)_2(OEt)^+ \rightarrow B(OMe)_2OH^+ + C_2H_4$
39.9	39.93	$B(OMe)(OEt)^+ \rightarrow B(OH)(OMe)^+ + C_2H_4$

[a] From Fallon and Lockhart [26].

alkyl borates. These particular rearrangements mentioned in Table 2.6 occur by ejection of olefin from the mixed ester (e.g., reaction 2.27). Again

$$H_2C-CH_2-O-B< \longrightarrow H_2C{=}CH_2 + H-O-B< \qquad 2.27$$

the presence of 20% of ^{10}B in natural boron complicates the spectra as does the presence of so many hydrogens. These systems are unsuitable for accurate equilibrium constant work. Accurate equilibrium data has been obtained in two labile reactions. The sensitivity of the instrument for nonisolable mixed compounds cannot of course be determined directly, but this drawback has been overcome in two ways: first, the sensitivity to the outer (unmixed) components is found, and that for the inner (mixed) components interpolated; second, the ion currents for parent ions which are related to concentration by

$$I = \sigma p \qquad 2.28$$

(where I is the ion current, σ is a constant, and p is partial pressure) can be obtained and the constant $K = I_{products}/I_{reagents}$ [which equals the true equilibrium constant times a constant factor $C = \sigma_{products}/\sigma_{reagents}$] is calculated at various temperatures. A plot of $\log K$ against $1/T$ provides a value of ΔH which is, of course, quite independent of sensitivity for each compound. The interpolative method has been used for arsenic halides [27] and the ion current method for boron halides [28].

Ionization potentials

Appearance potentials (P_A) of ions in the mass spectrum can be obtained from mass spectral data [29] and these are related to the strength of the bond broken in the step

parent → fragment ion   BX → B$^+$ + X·

by

$$P_{A(B^+)} = D_{(B-X)} + P_{I(B\cdot)} \qquad\qquad 2.29$$

where P_I refers to ionization potential. The bond dissociation energies $D_{(B-X)}$ obtained in this way are usually higher than those measured in other ways, however, their internal consistency is adequate to be useful in comparison of bond strengths in strongly nonrandom reactions.

IV. Polarography

A typical graph of current versus applied voltage (polarogram) during the reversible reduction of a metal ion at a dropping cathode is shown in Fig. 2.6. This indicates two features: (1) the half-wave potential $E_{1/2}$ which is

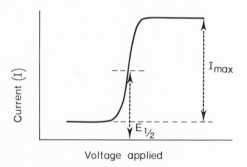

Fig. 2.6. Schematic representation of typical polarograph, showing I_{max} and $E_{1/2}$.

characteristic of the ion reduced but independent of its concentration, and (2) the height of the wave (limiting value of the current attained, I_{max}) which is proportional to concentration of the reducible ion. These two features make polarography an extremely useful technique (where applicable) for both qualitative and quantitative analysis of metal and other cations [30, 31]. Polarograms of mixtures of complexes

$$MX_J + MY_J \rightleftharpoons MX_{(J-1)}Y, \qquad \text{etc.}$$

can sometimes be used to show the occurrence of redistribution, the rate of reaction being relevant. For a reversible system in which redistribution is slow in comparison to the time required for measurement of the polarogram, a wave can be seen for each complex MX_j, MY_j, $MX_{j-1}Y$, etc. (unless these dissociate rapidly, $MX_j \rightarrow M + jX$, etc., when the diffusion of M is rate determining). For such reactions the appearance of a mixed product has been deduced from the appearance of a new wave associated with this species. Since the height of the new wave is proportional to concentration of the new species, the time dependence of wave height could be used to give rate constants. For the rapid reactions (relative to the time required to obtain the polarogram) only an averaged wave would be seen.

Thermodynamic data can also be obtained from polarography. The half-wave potential $E_{1/2}$ of a complex ion MX_j differs from that of a simple solvated metal ion M_{aq} and in favorable circumstances, a direct measurement of the difference in half-wave potential $\Delta(E_{1/2, \text{metal}} \sim E_{1/2, \text{complex}})$ will lead to the stability constant, β_j for the complex (where the reductions are reversible). The relationship between Δ and β is given by

$$\Delta(E_{1/2, \text{metal}} - E_{1/2, \text{complex}}) = (0.0591/n) \log_{10} \beta_j$$
$$+ j(0.0591/n) \log_{10} [X] \qquad \text{(at 25°)} \qquad 2.30$$

where n is the number of electrons required in the reduction, β is the stability constant of the complex MX_j and [X] is the concentration of X, if activities are ignored. The coordination number j can be found from the (linear) plot of $\log_{10}[X]$ against the half-wave potential of the complex [31]. Where there is more than one stable complex at any particular ligand concentration, e.g., MX_j, MX_{j-1} (step equilibria), the plot of $\log_{10}[X]$ against $E_{1/2}$ complex is usually continuous unless the individual values of β differ by at least a power of 10. Deford and Hume [32] devised a method to determine the individual β's, and the method has been extended by Schaap and McMasters [33] to deal with systems containing more than one type of ligand. Graphical solutions are possible for both these methods, and the resulting β values in appropriate ratio give the equilibrium constant K for the redistribution involved (reaction 2.31).

$$MX_j + MY_j \rightleftharpoons MX_{j-1}Y, \text{ etc.,} \qquad M + nX + MY \rightarrow MX_{j-1}Y, \text{ etc.} \qquad 2.31$$

A square-planar complex of nickel containing two bidentate ligands per molecule III was found to exchange ligands when mixed with another similar nickel complex and an indication of the rate of the process was obtained using polarography [34]. An equimolar mixture of the two dithiolates,

$$RC-S \diagdown \diagup S-CR'$$
$$\qquad Ni$$
$$RC-S \diagup \diagdown S-CR'$$

(a) $R = R' = CF_3$
(b) $R = R' = CN^-$
(c) $R = CN^-, R' = CF_3$

III

IIIa and IIIb initially had two reduction waves, one for each complex. With time a third wave at an intermediate $E_{1/2}$ appeared, representing the mixed complex IIIc. Equilibrium between the three dithiolates IIIa, b, and c was reached after 37 days at 40° in methylene chloride solvent.

An example of a thermodynamic study of mixed-ligand complexes using polarography is provided by the work of Schaap and McMasters [33] on the copper(II)–ethylenediamine–oxalate system in water. In this study the complex formation and redistribution reactions are rapid so that individual reduction waves for each possible complex $Cu(en)_2^{2+}$, $Cu(ox)_2^{2-}$ and $Cu(en)(ox)$, $Cu(en)^{2+}$, and $Cu(ox)$ are not seen. The variation of $E_{1/2}$ as the concentration of each ligand is independently varied yields a value for the formation constant for each species present. It is useful to note the conversion from formation constant data to randomization or redistribution constant (K_r). The formation constant K_{nm} for species $Cu(en)_n(ox)_m$ is given by

$$K_{nm} = [Cu(en)_n(ox)_m]/[Cu][en]^n[ox]^m \qquad 2.32$$

where n or m can be 0, 1, 2 and $n = 2 - m$. The redistribution constant K_r

$$Cu(en)_2^{2+} + Cu(ox)_2^{2-} \rightleftharpoons 2Cu(en)(ox) \qquad 2.33$$

for reaction (2.33) is given by

$$K_r = [Cu(en)(ox)]^2/[Cu(en)_2^{2+}][Cu(ox)_2^{2-}] = K_{11}/K_{20}K_{02} \qquad 2.34$$

V. Gas Chromatography

This technique has not been used much in redistribution work, but it has great potential for the separation of mixed products, particularly where there is very little difference in their boiling points. The principle is that materials are separated not only by their difference in volatility but also by differential absorption [35]. The mixture to be separated is mixed with carrier gas and passed through a stationary phase consisting of any suitable non-volatile liquid on a supporting solid column. Detection of the effluent gas depends on difference in physical properties between carrier gas and effluent gas, e.g., thermal conductivity, gas density, flame temperature of burning exit gases, etc. The order in which materials come off the column (order of retention times) is determined by absorption characteristics and calibration of a particular column for each component of the mixture to be separated is performed. The method also enables concentration to be determined from peak area; thus it can be used for quantitative estimation as well as mere detection of mixed products. The procedure can be both rapid and quantitative, but is not at present suitable for rapid redistribution reactions. Conceivably the method could be applied to exchanging systems where exchange life times were considerably shorter than retention times but such developments have not yet been made. (Other chromatographic techniques, e.g., paper or thin film, might be suitable separation methods for redistribution mixtures, but not normally quantitative and certainly not rapid. They are unlikely to find much application.)

Redistributing systems of carbonyl and trifluorophosphine on molybdenum have been separated by preparative gas chromatography using 15% Dow-Corning Silicone diffusion pump fluid on 30/60 mesh Kromat FB. Retention time progressively decreases as the amount of PF_3 substitution increases and the parent hexacarbonyl shows the longest retention time.

$$2Mo(PF_3)(CO)_5 \rightleftharpoons Mo(CO)_6 + Mo(PF_3)_2(CO)_4 \quad \text{(cis and trans)} \quad 2.35$$
relative retention times:
0.6   1   0.4   0.34

The peaks are so well separated for Eq. 2.35 that 20 μliter (20 nm^3) of the neat liquid could be separated on the column in one run [36].

Ester group interchange on silicon has also been investigated by gas chromatography. A liquid phase of Tween 80 (10%) was used on a solid C-22 firebrick support, and the concentration of each ester determined by electronic integration. The retention times of $(EtO)_4Si$ and $(MeO)_4Si$ were obtained for pure samples and those of the mixed esters were intermediate as anticipated. Individual ester concentrations could be determined to within 1% in the equilibrium mixture (equilibria 2.36) and so concentration

$$(MeO)_4Si + (EtO)_4Si \rightleftharpoons (MeO)(EtO)_3Si + (MeO)_3(EtO)Si \quad 2.36$$
$$2(MeO)_2(EtO)_2Si$$

equilibrium constants could be calculated for the three independent equilibria inherent in the reaction (2.36).

The reaction is quite slow at 150° (<4 days) and the column temperature used in the separation was 80°C, so there was no fear of scrambling in the column [37].

VI. Phase Diagrams

Use can be made of solid, liquid, and vapor equilibria in studies of re-distribution. Solid–liquid and liquid–vapor phase diagrams are useful in indicating the presence of new species in a mixture, and can indeed be used to determine the extent of reaction in some cases. The systems most favorable would be those in which each component could be isolated and its physical properties (e.g., freezing point, cryoscopic constant, etc.) determined. For labile systems this is not possible, but the presence of additional species may still be obvious from the phase diagram. Conspicuous failures of this method occur where a series of solid solutions is obtained from mixtures, as in the phenylboron halides [38] and where a high degree of supercooling is observed for components (for phenyltin trichloride in an abortive investigation, the author found a metastable liquid phase persisted 40° below the freezing point). Some of the simpler phase equilibria which can be encountered for redistribution reactions are given in most undergraduate texts in physical chemistry [39]. The apparatus required for this technique is often widely available, and the technique so positively venerable and familiar, that it deserves more consideration. It should not be neglected in favor of more expensive and sophisticated techniques without good reason. For a more complete treatment, specialist texts such as that by Ricci [40] should be consulted.

VII. Use of Labels

Isotopic or stereochemical labels have been useful in establishing the occurrence of redistribution reactions. Isotopic labels are necessary for reactions in which no net chemical change occurs (e.g., Eq. 2.37). In this example,

$$C^{35}Cl_4 + C^{37}Cl_4 \; \rightleftharpoons \; C^{35}Cl_3{}^{37}Cl + C^{35}Cl_2{}^{37}Cl_2, \qquad \text{etc.} \qquad 2.37$$

mass differences might be detected by mass spectrum. Radioactive tracers are also convenient, as shown in equilibrium 2.38. Stereochemical labels

$$BCl_3 + POCl_3{}^* \; \rightleftharpoons \; BCl_3{}^* + POCl_3 \qquad 2.38$$

are used mostly in determining the steric course of the reaction as for instance in the reduction of boron trichloride to diborane with an optically active silane (Eq. 2.39) [41]. The resultant chlorosilane has the same absolute

$$MePh\text{-}\alpha\text{-}NpSiH + BCl_3 \; \rightleftharpoons \; MePh\text{-}\alpha\text{-}NpSiCl + B_2H_6, \qquad \text{etc.} \qquad 2.39$$

configuration as the starting silane, so we may infer a quasi-cyclic disposition of reagents during reaction (flank attack on silicon as shown in structure IV). Silicon must be five coordinate, although the precise timing of bond-making and -breaking steps is unknown. Sometimes a technique of double labeling is more powerful than a single label. A sample of optically resolved *sec*-butylmercuric bromide, labeled with radiotracer mercury (^{203}Hg) was used by Charman *et al.* [42] to investigate the mechanism of reaction 2.40. Here

$$sec\text{-}Bu_2Hg + sec\text{-}Bu^*Hg'Br \rightleftharpoons sec\text{-}Bu^*Hg'\text{-}sec\text{-}Bu + sec\text{-}BuHgBr \qquad 2.40$$

the stereochemical label indicates retention of configuration during transfer from radiomercury and the rate of alkyl exchange can be measured through polarimetry, while the rate of mercury exchange can be obtained radio-metrically from the radioactive label. The two rates are found to be equal, proving that the exchange goes through a stage in which two mercury atoms are attached simultaneously to the optically active carbon (V) (front–side attack). Many inorganic compounds feature stereoisomers which can be

IV   V

distinguished by physical methods other than optical activity and provide suitable labels, e.g., the two configurations of vinyl groups attached to trivinylboron (seagull and swastika, distinguishable by electron spectra) [43] or the stereochemically nonequivalent methylene hydrogens of a stable *neo*-hexyl metal compound (these produce an AA'BB' pattern in ^{1}H NMR spectra, whereas A_2B_2 patterns result for labile *neo*-hexyl metal bonds) could be used to study mechanism [44].

REFERENCES

1. M. L. Delwaulle, M. B. Buisset, and M. Delhaye, *J. Amer. Chem. Soc.* **74**, 5768 (1952).
2. A. Finch, I. J. Hyams, and D. Steele, *J. Mol. Spectrosc.* **16**, 103 (1965).
3. A. J. Banister, N. N. Greenwood, B. P. Straughan, and J. Walker, *J. Chem. Soc.* **1964**, 995.
4. W. J. Lehmann, H. G. Weiss, and I. Shapiro, *J. Chem. Phys.* **30**, 1226 (1959).
5. T. L. Cottrell, "The Strengths of Chemical Bonds." Butterworth, London and Washington, D.C., 1954.
6. G. Herzberg, "Molecular Spectra and Molecular Structure," Chapter 5, p. 501. Van Nostrand, Princeton, New Jersey, 1945.

7. H. C. Urey, *J. Chem. Soc.* **1947**, 562.
8. K. B. Wiberg, "Physical Organic Chemistry." Wiley, New York, 1964.
9. K. S. Pitzer, "Quantum Chemistry." Constable, London, 1953.
10. G. J. Janz, *Quart. Rev. Chem. Soc.* **9**, 229 (1955).
11. J. Jakes and D. Papousek, *Collect. Czech. Chem. Commun.* **26**, 2110 (1961).
12. G. Nagajaran, *Bull. Soc. Chim. Belg.* **71**, 65, 73 (1962).
13. T. G. Spiro and D. N. Hume, *J. Amer. Chem. Soc.* **83**, 4305 (1961).
14. F. J. C. Rossotti and H. Rossotti, "Determination of Stability Constants and other Equilibrium Constants in Solution." McGraw-Hill, New York, 1961.
15. M. M. Crutchfield, C. H. Dungan, J. H. Letcher, V. Mark, and J. R. Van Wazer, "Topics in Phosphorus Chemistry," Vol. 5. Wiley (Interscience), New York, 1967.
16. A. Allerhand, H. S. Gutowsky, J. Jonas, and R. A. Meinzer, *J. Amer. Chem. Soc.* **88**, 3185 (1966).
17. N. S. Ham and T. Mole, *Progr. Nucl. Resonance Spectrosc.* **4**, 91 (1969).
18. J. J. Burke and P. C. Lauterbur, *J. Amer. Chem. Soc.* **83**, 326 (1961).
19. K. Moedritzer, *Advan. Organometal. Chem.* **6**, 171 (1968).
20. K. Moedritzer and J. R. Van Wazer, *J. Amer. Chem. Soc.* **86**, 802 (1964).
21. D. Forster, K. Moedritzer, and J. R. Van Wazer, *Inorg. Chem.* **7**, 1138 (1968).
22. A. Chakravorty and R. H. Holm, *J. Amer. Chem. Soc.* **86**, 3999 (1964).
23. G. Calingaert, F. W. Lamb, and F. Meyer, *J. Amer. Chem. Soc.* **71**, 3709 (1949).
24. P. M. Boorman, N. N. Greenwood, and H. J. Whitfield, *J. Chem. Soc. A* **1968**, 2256.
25. R. E. Dessy, F. Kaplan, G. R. Coe, and R. M. Salinger, *J. Amer. Chem. Soc.* **85**, 1191 (1963).
26. P. J. Fallon and J. C. Lockhart, *Int. J. Mass Spectrosc. Ion Phys.* **2**, 247 (1969).
27. J. K. Ruff and G. Paulett, *Inorg. Chem.* **3**, 998 (1964).
28. R. F. Porter, D. R. Bidinosti, and K. F. Watterston, *J. Chem. Phys.* **36**, 2104 (1962).
29. R. I. Reed, *Quart. Rev. Chem. Soc.* **20**, 527 (1966).
30. I. M. Kolthoff and J. J. Lingane, "Polarography." Wiley (Interscience), New York, 1941.
31. D. R. Crow and J. V. Westwood, *Quart. Rev. Chem. Soc.* **19**, 57 (1965).
32. D. D. Deford and D. N. Hume, *J. Amer. Chem. Soc.* **73**, 5321 (1951).
33. W. B. Schaap and D. L. McMasters, *J. Amer. Chem. Soc.* **83**, 4699 (1961).
34. A. Davison, J. A. McCleverty, E. T. Shawl, and E. J. Wharton, *J. Amer. Chem. Soc.* **89**, 830 (1967).
35. A. B. Littlewood, "Gas Chromatography," 2nd ed. Academic Press, New York, 1970.
36. R. J. Clark and P. I. Hoberman, *Inorg. Chem.* **4**, 1771 (1965).
37. J. R. Van Wazer and K. Moedritzer, *Inorg. Chem.* **3**, 268 (1964).
38. J. C. Lockhart and A. Finch, unpublished work.
39. S. Glasstone, "Textbook of Physical Chemistry," 2nd ed. Van Nostrand, Princeton, New Jersey, 1946.
40. J. E. Ricci, "The Phase Rule and Heterogeneous Equilibria." Van Nostrand, Princeton, New Jersey, 1951.
41. C. J. Attridge, R. N. Hazeldine, and M. J. Newlands, *Chem. Commun.* **1966**, 911.
42. H. B. Charman, E. D. Hughes, C. K. Ingold, and F. G. Thorpe, *J. Chem. Soc.* **1961**, 1121.
43. D. R. Armstrong and P. G. Perkins, *Theor. Chim. Acta* **5**, 11 (1966).
44. M. Witanowski and J. D. Roberts, *J. Amer. Chem. Soc.* **88**, 737 (1966).

PART II

3

Group I

I. Lithium Alkyls, Aryls, and Silyls

Although the structure of alkyllithium compounds is not fully understood in all its aspects, exchange phenomena involving these entities have been discovered and can be convincingly explained in terms of intra- and inter-molecular processes. Since lithium alkyls are particularly effective catalysts for anionic polymerization of olefins [1], the occurrence of inter- and intramolecular processes in the alkyls themselves is of prime significance offering the possibility of tailoring catalysts to be of required activity.

The structures of alkyl- and aryllithiums are unusual; they are undoubtedly aggregated under normal conditions, although transient kinetically active monomers are also indicated. Dimer, tetramer, and hexamer aggregates are substantiated as, for example,

$(PhLi)_2$	$(Me_3SiCH_2Li)_4$	$(EtLi)_6$
dimer in ether	tetramer in benzene	hexamer in benzene
four-center bonds	four-center bonds	four-center bonds

The simplest postulate of bonding considers delocalization over three-center (or four-center) orbitals encompassing two (or three) lithium nuclei and one alkyl group [2]:

$$
\begin{array}{c}
\text{Li} \diagdown \quad \diagup \text{Li} \\
\text{R} \\
| \\
\text{Li}
\end{array}
$$

31

The unit in tetramer and hexamer can be thought of as a trigonal array of lithium atoms (triangular face) with the alkyl group R above the center of the face. A suitable combination of one orbital from each lithium and the alkyl group gives an appropriate bonding four-center orbital. Each trigonal unit is linked at the edges to other units, four faces forming a tetrahedron, six an octahedron with two unalkylated faces.

The system t-butyllithium–lithiomethyltrimethylsilane [3] illustrates well the occurrence of slow intermolecular exchange of lithium, with a half-life of 6–8 hr in cyclopentane. The ^7Li resonances of the two starting materials at room temperature are singlets 24.7 Hz apart; in mixtures, a slow reaction can be observed through the growth of new lithium signals at the expense of the original ones, and in equilibrium mixtures, five signals including the original two are seen. These can be accounted for as the five mixed species derived from tetramer aggregates* a–e

Li_4-t-Bu_4 Li_4-t-$Bu_3(CH_2SiMe_3)$ Li_4-t-$Bu_2(CH_2SiMe_3)_2$ Li_4-t-$Bu(CH_2SiMe_3)_3$
   a   b   c   d

$Li_4(CH_2SiMe_3)_4$
e

Since the lithium in each mixed tetramer should have more than one possible environment of alkyl groups (one group on each neighbor face, three in all), an assumption is necessary to explain observation of five rather than the eight possible signals. The four environments A–D are possible (R and R′ represent different groups):

If, within each tetramer, the environment is rapidly averaged by some intramolecular switching of alkyl groups, then five signals could be observed, one for each of the species a–e. This hypothesis of intramolecular switching is neatly substantiated at low temperatures when the lithium spectrum is again altered—only four lines are seen at the available resolution. None of them corresponds in chemical shift to the individual resonance of a or e. Instead each line corresponds to one of the four environments A–D rather

---

* The hexamer should show at least seven lines for slow intermolecular exchange. This does not take account of possible cis–trans isomers, which might well show separate lithium signals.

than to the individual species a–e. At low temperatures, the switching process has become sufficiently slow for the immediate environment of three alkyl groups to determine the chemical shift of any lithium atom. At higher resolutions, it is to be expected that each of the four lines would be appropriately split (Table 3.1). Peak areas of the four low-temperature signals

TABLE 3.1

ENVIRONMENT OF LITHIUM IN THE MIXED COMPOUNDS
a–e AS DETERMINED BY THE THREE ADJACENT ALKYL
GROUPS

A	B	C	D
aA	bB	cC	dD
bA	cB	dC	eD

are in reasonable agreement with those calculated on an assumption of random scrambling of alkyl groups on the tetramers, and the presence in line A of components aA, bA, etc., as indicated in Table 3.1. The intramolecular exchange must involve some molecular deformation, for example, to a planar intermediate (Eq. 3.1). In toluene, it is a factor of

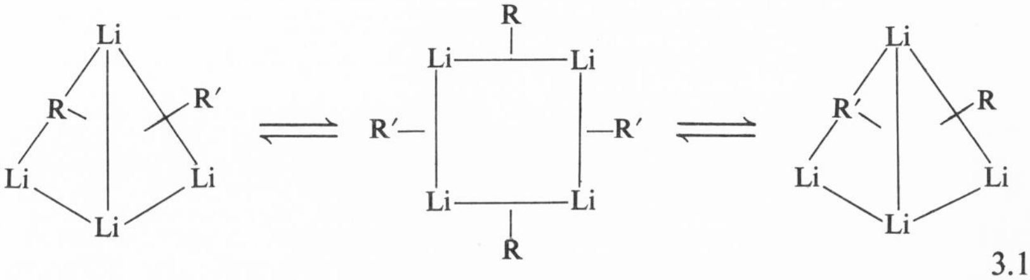

3.1

10^6 faster than the intermolecular reaction. The intermolecular reaction (scrambling of alkyl groups on lithium) may occur with a rate-determining dissociation to dimers (Eq. 3.2) of the butyl compound since the ^{7}Li signals

$$(\text{Li-}t\text{-Bu})_4 \ \rightleftharpoons \ 2(\text{Li-}t\text{-Bu})_2 \qquad\qquad 3.2$$

for silane-rich mixed compounds c and d appear earlier in the course of the reaction than that for b. The data yield a rate constant for this dissociation (Eq. 3.2); it is interesting that the production of dimer is considerably slower than the initiation step in anionic polymerizations using t-BuLi.

Spin coupling $^{13}C-^{7}Li$ has been observed in the ^{7}Li NMR spectrum of *t*-butyllithium enriched with ^{13}C, in cyclohexane and toluene at -20 to $+80°$ [4]. This confirms that intermolecular exchange of lithium in hydrocarbon solvent is slow on the NMR time scale as postulated by Hartwell and Brown [3]. The ratio of intensities of lines in the observed spectrum has been measured but is not sufficient to distinguish between the spectrum anticipated for the slow intramolecular exchange favored by Hartwell and Brown [3] (coupling of ^{7}Li to the immediate environment of up to three ^{13}C nuclei should give rise to seven lines) and that anticipated for the rapid exchange limit (coupling of ^{7}Li to between zero and four ^{13}C nuclei should give a nine-line pattern for a random sorting of ^{12}C and ^{13}C). This study also indicated rapid intermolecular exchange for *n*-butyllithium, since no $^{13}C-^{7}Li$ coupling was seen.

$$M—\underset{\underset{H_A}{|}}{\overset{\overset{H_A}{|}}{C}}—\underset{\underset{H_B}{|}}{\overset{\overset{H_B}{|}}{C}}—\underset{\underset{CH_3}{|}}{\overset{\overset{CH_3}{|}}{C}}—CH_3$$

I

Neo-hexyllithium in ether provides an insight into the stereochemistry of processes in these systems [5]. The ^{1}H NMR spectrum of the α-methylene group of a configurationally stable neohexyl (see structure I) has an AA'BB' pattern, but if there is rapid configurational exchange (see Eq. 3.3)

$$\begin{array}{c} CMe_3 \\ H_A\!-\!\!\!+\!\!\!-\!H_{A'} \\ H_B\!-\!\!\!+\!\!\!-\!H_{B'} \\ M \end{array} \quad \rightleftharpoons \quad \begin{array}{c} CMe_3 \\ H_{A'}\!-\!\!\!+\!\!\!-\!H_A \\ H_{B'}\!-\!\!\!+\!\!\!-\!H_{B'} \\ M \end{array} \qquad 3.3$$

at the α-carbon, a simple A_2B_2 pattern emerges. Dilute ether solutions of neohexyllithium display the A_2B_2 pattern and the activation energy for the (first-order) inversion is found to be 62.8 kJ mole^{-1} (Table 3.2).

For neohexyl derivatives of other metals, inversion occurs with different activation energy, thus a dissociative step (ionic) is suggested (Eq. 3.4).

$$Li_4R_4 \rightleftharpoons Li_4R_3^+ + R^- \qquad 3.4$$
$$\text{carbanion}$$

However, the activation energy is of the same order as that for dissociation of alkyllithium tetramers to dimers which is not ionic. What is needful is a

TABLE 3.2

ACTIVATION ENERGIES FOR SOME REACTIONS OF ALKYLLITHIUMS

Process	Solvent	$E^{\ddagger}$ (kJ mole^{-1})
$(MeLi)_4 \rightarrow 2(MeLi)_2$	Ether	46.0 [6]
Neohexyl-Li inversion	Ether	62.8 ± 8.4 [5]
$(Li\text{-}t\text{-}Bu)_4 \rightarrow 2(Li\text{-}t\text{-}Bu)_2$	Cyclopentane	100.3 ± 16.6 [4]

study of exchange of neohexyllithium with another suitable alkyllithium of considerably different lithium chemical shift, to determine the relation between such processes as are shown in equilibria 3.1–3.3

Random exchange of methyl and ethyl groups between methyllithium and ethyllithium (probably present as tetramers) in ether solution can be detected by ^{7}Li and ^{1}H NMR and is thought to occur through dissociation to dimers and reassociation [6]. It is noted that ethyl scrambling is more rapid than methyl. Activation energy is given in Table 3.2 for the dissociation step. The mixed systems phenyllithium (dimer) and methyl- or ethyllithium (tetramer) have been similarly studied. Here the reagents are in different states of aggregation in ether, and so the states of aggregation in the mixed species are of great interest. Two mixed species, probably Li_4Et_3Ph and a 1:1 complex $(Li_2EtPh)_n$, can be detected by ^{7}Li resonance, but by an unhappy accident, the ^{7}Li chemical shift of methyllithium is the same as that of phenyllithium. Therefore, for methyl mixtures only ^{1}H resonance work is applicable; two new methyl resonances are observed in mixtures and are thought to be Li_4Me_3Ph and $(Li_2MePh)_n$ but the exact degree of aggregation is not certain. It does seem that phenyl and alkyl groups (Me or Et) do not scramble randomly on lithium. Phenyl groups show a preference for bonding in the $(Li_2RPh)_n$ entity rather than on the electron deficient tetramer structure Li_4R_3Ph [7].

Closely similar studies use ^{7}Li and ^{1}H NMR techniques on ether solutions of lithium alkyls or aryls with aryls or alkyls of Group II metals (Zn, Mg) and Group III elements (B, Al, Ga). Considerable information is available as to the rates and mechanisms of the inherent exchange reactions and the work incidentally confirms or establishes the structures of some "ate" complexes in solution. Some activation energies are given in Table 3.3. In all the Group III systems LiR–LiMR$_4$ (where R is alkyl, M is element) bar those of aluminum, exchange of lithium is rapid, but exchange of alkyl

TABLE 3.3

Activation Energies for Exchange Reactions Occurring in Mixtures of Lithium Alkyl/Aryl with Alkyls and Aryls of Groups II and III

Reaction	Exchange followed	$E^{\ddagger}$ (kJ mole^{-1})	Ref.
$AlMe_3 + LiAlMe_4$	Me	38 ± 4.2	8
$(LiMe)_4 + LiAlMe_4$	Li	46.9 ± 4.2	8
$(LiMe)_4 + LiBMe_4$	Li	46.9 ± 4.2	8
$(LiMe)_4 + Li_2ZnMe_4$	Li	45.6 ± 12.6	10
$(LiMe)_4 + Li_3ZnMe_5$	Me	35.5 ± 12.6	10
$(LiMe)_4 + Li_2MgMe_4$	Li	43.1 ± 12.6	10
$(LiMe)_4 + Li_2MgMe_4$	Me	42 ± 12.6	10
$Me_2Mg + Li_2MgMe_4$	Me	62.7 ± 12.6	10
$(PhLi)_2 + Li_2ZnPh_4$	Li	49.4 ± 4.2	11
$(PhLi)_2 + Li_2MgPh_4$	Li	51.5 ± 4.2	11

groups is slow [8]. For alkyl group exchange in the methyl–aluminum system $AlMe_3$–$LiAlMe_4$, the rate-determining step is possibly formation of a solvent-separated ion pair (Eq. 3.5) and subsequent rapid exchange of

$$LiAlMe_4 \rightleftharpoons Li^+\|AlMe_4^- \text{ ion pair} \qquad 3.5$$

methyl between ionic $AlMe_4^-$ and $AlMe_3$. (See Table 3.3 for activation energy.) Lithium exchange occurs at the same rate in $LiBMe_4$ or $LiAlMe_4$ mixtures with Li_4Me_4 and it is reasonable to assume the same mechanism, with rate-determining dissociation of the common component (Eq. 3.6).

$$Li_4Me_4 \rightleftharpoons 2(LiMe)_2 \qquad 3.6$$

$$LiMMe_4 \rightleftharpoons Li^+| \, |MMe_4^- \qquad 3.7$$

Further rapid exchange of dimer with a solvent-separated lithium ion could follow via a transition state such as II, in which all lithium atoms are equivalent (Eqs. 3.7 and 3.8). The postulated $(Li_nR_{n-1})^+$ ions such as II are known to be the major species produced on electron impact from Li_4R_4 [9], although the structure is mere guesswork. Since ethyllithium dissociates to dimer more readily than methyllithium, (Eq. 3.6), this mechanism leads us to expect more rapid reaction in related ethyl systems. The ethyl mixtures $LiEt–LiAlEt_4$ and $LiEt–LiGaEt_4$ do in fact exchange lithium somewhat faster than the methyl systems just discussed. Coupling of methyl protons with B or Al is observed at the highest temperatures studied ($+50°$) and thus dissociation of MMe_3 groups from the MMe_4^- entity is ruled out as a possible exchange route:

$$LiMMe_4 \quad \xrightarrow{\quad | \quad} \quad LiMe + MMe_3$$

"Ate" complexes of the type Li_2MR_4 and Li_3MR_5 (M is Zn or Mg) are formed at high ($>2:1$) molar ratios of lithium to magnesium or zinc alkyl, again in ether solutions and the zinc "ate" complexes are thermodynamically more stable. Lithium and alkyl exchange appear to have the same rate here and that rate is greater in the zinc than in the magnesium system [10], which suggests the complex Li_2MR_4, etc., is involved in the rate-determining step.

The mechanisms are thought to be different from those for the Group III systems just discussed. The 2:1 and 3:1 complexes exchange more rapidly with each other than they do with alkyllithium, and a rapid transfer of one monomeric methyllithium unit between the two is suggested (Eq. 3.9).

$$Li_2MMe_4 + Li_3MMe_5 \;\rightleftharpoons\; Li_2MMe_4 \cdots LiMe \cdots Li_2MMe_4 \qquad 3.9$$
$$\text{III}$$

The stereochemistry of the intermediate III is not known. Since the two complexes effectively become intermediate III under fast exchange conditions, it is possible to consider the overall exchange (of lithium and of methyl) as between intermediate III and methyllithium. The activation energies (Table 3.3) are the same for this process as for dissociation of methyllithium tetramer (Table 3.2), which is a probable rate-determining step (Eq. 3.6), the dimer then exchanging lithium and alkyl with intermediate III (Eq. 3.10). The striking feature of these Group II systems, as opposed to the Group III systems, is that alkyl exchange is as fast as lithium exchange, necessitating using equilibrium 3.10 (Group II) and quite different reactions of the dimer $(MeLi)_2$ equilibrium 3.8 (for Group III). When the ratio of

$$(LiMe)_2{}^* + \text{III} \;\rightleftharpoons\; (LiMe)_2 + \text{III}^* \qquad 3.10$$

lithium to magnesium is less than 2, the mixture consists of methylmagnes-
ium and the 2:1 complex Li_2MgMe_4. Alkyl transfer between the two
(Eq. 3.11) has an activation energy of 62.8 kJ mole^{-1}. The corresponding
zinc system is probably in a state of rapid exchange at $-107°$, for even at
this temperature it was not possible to detect a signal for free $ZnMe_2$.

$$Me_2^*Mg + Li_2MgMe_4 \rightleftharpoons Me_2Mg + Li_2MgMe_4^* \qquad 3.11$$

By contrast, in the phenyl–lithium–magnesium (or zinc) system, only the
2:1 and 1:1 complexes Li_2MPh_4 and $LiMPh_3$ are detected, and the phenyl
group exchange is slower than lithium exchange between the 2:1 complex
and phenyllithium. Different mechanisms have been proposed [11]. The
rate of lithium exchange is independent of the concentration of aryllithium
and the aryllithium must be involved *after* the rate-determining step, which
must be some dissociation of the 2:1 complex, probably to an ion pair.

$$Li_2MPh_4 \rightleftharpoons Li^+ \parallel LiMPh_4^- \qquad \text{rate determining} \qquad 3.12$$

$$(Ph^*Li)_2 + Li^+ \rightleftharpoons Li^* + PhLiPhLi^* \qquad 3.13$$

This is reminiscent of the processes in Group III (3.7 and 3.8) except that
the aryllithium is present as dimer and the preliminary dissociative step
(3.6) is not necessary. The proton spectra of the aromatic ring are very
complex and their behavior at various temperatures does not permit any
simple analysis to give an activation energy, but the rate of phenyl group
exchange is seen to be independent of $(PhLi)_2$ concentration and possibly
involves molecular dissociation of the 2:1 complex (Eq. 3.14) which is
assumed to be slower than the heterolysis (Eq. 3.12). Phenyl exchange is
slower for zinc than for magnesium, and, consistent with this mechanism,
the zinc complex is more stable thermodynamically than the magnesium.

$$Li_2MPh_4 \rightleftharpoons Ph_2M + (Ph^*Li)_2 \qquad \text{rate determining} \qquad 3.14$$

$$(Ph^*Li)_2 + (PhLi)_2 \rightleftharpoons 2Ph^*LiPhLi \qquad 3.15$$

Distribution of alkyl and phenyl groups on magnesium and lithium in
ether has been examined. Mixed 2:1 complexes occur, $(Li_2MgMe_{4-n}Ph_n)$
and phenyl groups are found preferentially bonded in these rather than in
the lithium tetramer Li_4Me_3Ph [11].

There is a brief report that methyl exchange in dimethylmercury in ethers
(normally very slow) [12] is accelerated by addition of methyllithium
although no "ate" complex $LiHgMe_3$ is actually observed [13]. Since
methyl groups on mercury are still exchanging when those on lithium

are not, the exchange may take place through catalytic amounts of LiHgMe$_3$:

$$LiHgMe_3* + HgMe_2 \rightleftharpoons LiHgMe_3 + HgMe_2*$$

Brightly colored "ate" complexes are produced in the trimethylsilyl-mercury series when trimethylsilyllithium is added, LiHg(SiMe$_3$)$_3$ and Li$_2$Hg(SiMe$_3$)$_4$. The 1:1 complex undergoes self-exchange, which is considerably accelerated by excess Hg(SiMe$_3$)$_2$ in dimethoxyethane solution. The 2:1 complex undergoes self-exchange via the reaction

$$Li_2Hg(SiMe_3)_4 \rightleftharpoons LiSiMe_3 + LiHg(SiMe_3)_3$$

The equilibrium lies to the left at low temperatures and the right at room temperatures [14].

Scrambling of lithium and bromine on alkyl, alkenyl, and aryl residues has been investigated, and the material is included in the section on lithium for comparison. Extensive kinetic and thermodynamic data are available for some systems and a comparison between salt-free† and salt-containing† lithium aryls has been made. Striking differences in reactivity result in salt-free systems, a factor of utmost importance to their performance as anionic polymerization catalysts.

Some interesting steric data have been interpreted mechanistically. Tripticyl bromide is completely shielded from rear attack by a carbanion [15] so that its lithiation by butyllithium must occur via front-side attack of butyl on bromo, and lithium on the incipient tripticyl carbanion (Eq. 3.16). Octyllithium, as prepared from optically resolved 2-iodooctane by

$$(tript)Br + LiBu \rightleftharpoons tript \cdots \substack{Br \\ \\ Li^{\delta+}} \cdots \substack{\delta- \\ \\} Bu \rightleftharpoons (tript)Li + BuBr$$

3.16

reaction with *sec*-butyllithium, is partly racemic and the extent of racemization depends on solvent, which suggests a solvent-separated carbanion in the reaction path [16].

Several halogen–lithium exchanges have been investigated thermodynamically in ether at low temperatures [17]. The equilibrium constants K observed for the stoichiometric equation

$$RLi + R'I \rightleftharpoons R'Li + RI, \qquad K_{obs} = [R'Li][RI]/[RLi][R'I]$$

† Salt-containing lithium aryls are made from aryl halide, $RX + 2Li \rightarrow RLi + LiX$, and the equimolar lithium halide has not been removed. Salt-free lithium aryls do not contain residual lithium halide.

are independent of the concentration of organolithium (Table 3.4), even although RLi is known to be not a monomer but dimer, tetramer, etc. If it

TABLE 3.4

EQUILIBRIUM CONSTANTS (GIVEN AS $\log K_{obs}$) FOR THE REACTION OF RLi WITH R'I[a, b]

R	R'	$\log K_{obs}$
Ph	Vinyl	2.41
cyclo-Pr	Ph	0.98
n-Pr	Et	0.38
i-Bu	n-Pr	0.71
Neopentyl	i-Bu	0.88

[a] From Applequist and O'Brien [17].
[b] Ether solvent at −70°C.

is assumed that random scrambling occurs, K observed should be independent of RLi concentration [17].

The position of the equilibrium 3.17 has been investigated in both direc-

$$PhLi + Br\langle\bigcirc\rangle Y \rightleftharpoons Y\langle\bigcirc\rangle Li + PhBr \qquad 3.17$$

tions for a series of substituted aryl groups [18]. The equilibrium constant K for Eq. 3.17 as written, is given by

$$K = \left[Y-\langle\bigcirc\rangle-Li\right][PhBr]/[PhLi]\left[Y-\langle\bigcirc\rangle-Br\right] \qquad 3.18$$

where PhLi is the total phenyllithium concentration, irrespective of its state of aggregation, and is found to be invariant with aryllithium concentration. This is surprising in view of the fact that lithium aryls can dimerise and also associate in salt-containing solutions with lithium halide, but would be consistent with random scrambling of each aryl group

$$\left(Ph \text{ and } Y-\langle\bigcirc\rangle\right)$$

among its possible environments [17]. For a series of substituted aryls the logarithm of the equilibrium constant K can be correlated with the Hammett substituent constant σ (values of σ are apparently taken from the review by Jaffé [19]) with a very high ρ value of 5.5 (Table 3.5):

$$\log K/K_0 = \sigma\rho$$

It is seen that aryls with electron-withdrawing substituents form the more stable lithium compounds and, as a corollary, the more stable carbanions.

TABLE 3.5

EQUILIBRIUM CONSTANTS K AND OVERALL RATE CONSTANTS K_{obs} AS IN EQS. 3.17 AND 3.18 FOR A SERIES OF SUBSTITUENTS Y[a]

Substituent	σ	$\log K$	$\log k_{obs}$
m-CF$_3$	+0.415	2.46	2.23
p-Cl	+0.226	1.72	0.73
p-Ph	+0.009	0.58	−0.17
m-CH$_3$	−0.069	−0.06	−0.47
p-CH$_3$	−0.70	−0.22	−0.72
		ρ 5.5	ρ 4.0

[a] From Winkler and Winkler [18, 20].

Alternatively, the aryl group with the more electron-donating substituent is more stable as the bromide. Reaction 3.17 (Y = CH$_3$) is almost thermoneutral in ether ($\Delta H = 0.71 \pm 0.21$ kJ mole^{-1}) and tetrahydrofuran ($\Delta H = 0.04 \pm 0.54$ kJ mole^{-1}) in the temperature range from −10° to 25°.

In salt-containing mixtures, the rate of exchange between phenyllithium and substituted aryl bromides is second order overall, first order in each reagent: the reverse reaction has the same kinetic order. The ratio of forward and reverse rate constants reproduces the equilibrium constant previously measured. The rate constants obey a linear free energy relationship (as do the equilibrium constants) with a high ρ value 4.0 (Table 3.5). The high ρ value shows that the aryllithium with more stable carbanion reacts more slowly (i.e., is a weaker nucleophile) while the aryl bromide with the electron withdrawing substituents is the stronger electrophile, which, mechanistically means that electron-withdrawing substituents on the aryl bromide and electron-donor substituents on the aryllithium are most favorable for fast reaction.

Salt-free mixtures are more complicated. The rate again is first order in each reagent, but the overall second-order rate constant for the reversible reaction, k_{obs} increases linearly with increasing initial concentrations. The rate is higher than for salt-containing aryllithiums and this effect was quantitatively assessed: addition of lithium bromide up to a mole ratio of 2:1 decreases the rate of interconversion of salt-free phenyllithium and beyond this ratio has no further effect. This suggests that the state of aggregation of phenyllithium (which depends on salt concentration) is important in determining the difference in rate of salt-free and salt-containing aryllithium reagents. The rate data best fit the sequence in which random scrambling of aryl groups between possible aggregates is a feature. The presence of salt breaks up the dimers and obviates this step [18, 20].

II. Alkali β-Diketonates

Alkali metal derivatives of various fluorinated β-diketones have been shown to be polymeric (at least trimeric) in the vapor phase. The trimeric aggregates can be detected in the mass spectrometer (MS9), giving rise to such ions as $M_3(hfac)_3{}^+$, $M_3(hfac)_2{}^+$, $M_2(hfac)_2{}^+$, etc., where M is an alkali metal and hfac stands for hexafluoroacetylacetone [21]. When mixtures containing different ligands or alkali metals are inserted into the instrument, the exchange of metals between ligands or ligands between metals is apparent (Eqs. 3.19 and 3.20). Attempts to observe this exchange for lithium complexes

$$Li_3(hfac)_3 + Li_3(tfac)_3 \rightleftharpoons Li_3(hfac)_2(tfac) + Li_3(hfac)(tfac)_2 \qquad 3.19$$

$$Li_3(tfac)_3 + K_3(tfac)_3 \rightleftharpoons Li_2K(tfac)_3 + LiK_2(tfac)_3 \qquad 3.20$$

in solution using ^{19}F, ^7Li, and ^1H NMR signals have been hampered by the insolubility of the compounds in nondonor solvents [22]. The β-diketonate derivatives of lithium are soluble in donor solvents such as ethers and dimethylformamide, but are probably present as monomers (the remaining coordination positions are probably occupied by solvent molecules), since only one chemical shift was observed for each β-diketone present in the mixtures. A ligand in the homotrimer would be expected to differ in chemical shift from the same ligand in a mixed trimer.

REFERENCES

1. C. E. H. Bawn and A. Ledwith, *Quart. Rev. Chem. Soc.* **16**, 361 (1962).
2. T. L. Brown, D. W. Dickerhoof, and D. A. Bafus, *J. Amer. Chem. Soc.* **84**, 1371 (1962).

3. G. E. Hartwell and T. L. Brown, *J. Amer. Chem. Soc.* **88**, 4625 (1966).
4. L. D. McKeever and R. Waack, *Chem. Commun.* **1969**, 750.
5. M. Witanowski and J. D. Roberts, *J. Amer. Chem. Soc.* **58**, 737 (1966).
6. L. M. Seitz and T. L. Brown, *J. Amer. Chem. Soc.* **88**, 2174 (1966).
7. L. M. Seitz and T. L. Brown, *J. Amer. Chem. Soc.* **89**, 1607 (1967).
8. K. C. Williams and T. L. Brown, *J. Amer. Chem. Soc.* **88**, 4134 (1966).
9. G. E. Hartwell and T. L. Brown, *Inorg. Chem.* **5**, 1257 (1966).
10. L. M. Seitz and T. L. Brown, *J. Amer. Chem. Soc.* **88**, 4140 (1966).
11. L. M. Seitz and T. L. Brown, *J. Amer. Chem. Soc.* **89**, 1602 (1967).
12. R. E. Dessy, F. Kaplan, G. R. Coe, and R. M. Salinger, *J. Amer. Chem. Soc.* **85**, 1191 (1963).
13. L. M. Seitz and S. D. Hall, *J. Organometal. Chem.* **15**, P7 (1968).
14. T. F. Schaaf and J. P. Oliver, *J. Amer. Chem. Soc.* **91**, 4327 (1969).
15. G. Wittig and U. Schollkopf, *Tetrahedron* **3**, 91 (1958).
16. R. L. Letsinger, *J. Amer. Chem. Soc.* **72**, 4842 (1950).
17. D. E. Applequist and D. F. O'Brien, *J. Amer. Chem. Soc.* **85**, 743 (1963).
18. H. J. S. Winkler and H. Winkler, *J. Amer. Chem. Soc.* **88**, 964 (1966).
19. H. H. Jaffé, *Chem. Rev.* **53**, 191 (1953).
20. H. J. S. Winkler and H. Winkler, *J. Amer. Chem. Soc.* **88**, 969 (1966).
21. J. R. Majer and R. Perry, *Chem. Commun.* **1969**, 271.
22. A. C. Linsdell and J. C. Lockhart, unpublished work.

4

Group II

In Group IIA, magnesium, principally because of its role in the Grignard reagent, is the only element whose redistribution reactions have been extensively studied. Considerable interest will attach to beryllium because of its simplicity (atomic number 4) but not many redistribution studies have been made as yet. The Group IIB metals have proved of great interest in redistribution work, the organomercurials in particular having provided much of the advanced mechanistic work on alkyl exchange reactions. Reactions of monofunctional substituents on mercury are simple to interpret since mercury is bifunctional. The tendency of zinc and cadmium to form aggregates accounts for some rather surprising new redistribution reactions in which the metal is no longer merely bifunctional. Exchanges between two different central atoms from Group II have been investigated kinetically and provide useful comparisons of metal with metal.

I. Beryllium

The complex ion $Be(DMF)_4^{2+}$ as its perchlorate reacts with beryllium acetylacetonate $Be(acac)_2$ in dimethylformamide (DMF) solutions to give the mixed ion (Eq. 4.1). Proton NMR signals for the dimethylformamide protons in the solvent and in the beryllium coordination sphere, and the

acetylacetonate groups can be resolved for the different species below 5°
[1]:

$$Be(DMF)_4^{2+} + Be(acac)_2 \rightleftharpoons 2(DMF)_2Be(acac)^+ \qquad 4.1$$

where $K = 10 \pm 2$, $\Delta H = 0 \pm 4$ kJ mole^{-1}, and $\Delta S = +21 \pm 13$ J °K^{-1}. The
low enthalpy is unexpected in view of the difference in formal charge on the
three entities in Eq. 4.1.

The occurrence of the Schlenk equilibrium 4.2 has been inferred in mix-

$$Ph_2Be + Be^*Br_2 \rightleftharpoons Ph_2Be \cdot BeBr_2 \xrightleftharpoons{\quad\quad} 2PhBeBr \qquad 4.2$$

tures of diphenylberyllium and beryllium bromide, since no exchange of
radioactive beryllium tracer (^{7}Be) from beryllium bromide to the diaryl
took place in 15 min in ether solution [2]. However, for dimethylberyllium
and beryllium dibromide mixtures in ether, methyl groups do exchange
rapidly between beryllium atoms with formation of monomeric methyl-
beryllium bromide [3] (4.3). Separate NMR signals are seen for the alkyl

$$Me_2Be + BeBr_2 \rightleftharpoons 2MeBeBr \qquad 4.3$$
$$a b$$

hydrogens on sites a and b in Eq. 4.3 at -75°C but these are coalesced at
35°C. The separation between the two signals is only 4 Hz at 60 MHz and
the ^{9}Be nuclear spin quantum number is $\frac{3}{2}$, so it is surprising that signals
can be distinguished at all.

II. Magnesium

Because of the interest in Grignard reagents and their structure, a spate of
publications has appeared on alkyl exchange on magnesium: the most
recent reviews are by Ashby [4] and Wakefield [5]. The Grignard reagent
apparently enters into equilibria of the type in Eq. 4.4 with the quantitative

$$(RMgX)_n \rightleftharpoons RMgX \rightleftharpoons R_2Mg + MgX_2 \rightleftharpoons R_2Mg \cdot MgX_2 \qquad 4.4$$

features determined by solvent, substituents, etc., so that it is impossible
to arrive at a single definite structure for the Grignard reagent. Rather it
embodies the labile system 4.4 where n may be a small integer. Ionic structures are also known in some circumstances. Monomeric species can exist
in diethyl ether, the time-hallowed solvent for Grignard reactions, but at
concentrations above 0.3 molar they are likely to be associated. In the basic

solvent triethylamine, the simple monomer RMgX is usually found [4], but in hydrocarbon solvents there are highly associated mixtures such as R_3Mg_2X [6]. The solid $PhMgBr \cdot 2Et_2O$ has been found to have tetrahedral coordination [7].

Calorimetric measurements have been made of the exothermic reaction of diethyl- or diphenylmagnesium with equimolar magnesium bromide or iodide in ether (Table 4.1). At concentrations of 0.1 M, i factors around 1 show that the product to a first approximation is monomeric (Table 4.2) [8].

TABLE 4.1

THERMODYNAMIC DATA FOR THE REACTION $R_2Mg + MgY_2$ IN ETHERS[a]

Solvent	M (mole liter^{-1})	R	Y	$-\Delta H$ (kJ mole^{-1})	ΔS (J K^{-1})
Ether	0.02	Et	Br	15.7	−1.3
Ether	0.2	Et	Br	14.0	—
THF	0.2	Et	Br	25.6	99
Ether	0.02	Ph	Br	8.6	—
Ether	0.1	Ph	Br	6.8	5.0
THF	0.2	Ph	Br	11.8	50.6

[a] From Smith and Becker [8, 9].

(The i factor for the corresponding Grignard reagent is the same.) Reaction in tetrahydrofuran solutions is endothermic but again the i factor of the product shows it to be a monomer and to coexist in equilibrium with

TABLE 4.2

i FACTORS OBSERVED FOR EQUIMOLAR
MIXTURES OF R_2Mg AND MgY_2 IN ETHER[a]

R	Y	Molarity	i
Et	Br	0.1	1.04
Et	Br	0.3	1.62
Ph	Br	0.1	1.12
Ph	Br	0.3	1.74

[a] From Smith and Becker [8].

dialkylmagnesium and magnesium halide [9]. The ethylmagnesium chloride mixture contains other species, notably $EtMg_2Cl_3$. Rather different entropy values (Table 4.1) for the alkyl exchange reaction are attributed to differential solvation effects on magnesium halide. The higher i factors observed at higher concentrations in ethers indicate associated species, and the obvious method of linkage is by halogen bridging.

Proton NMR spectra have not been particularly helpful in the Grignard system, possibly because chemical shift differences between alkyl groups attached to magnesium (R_2Mg and $RMgX$) are too slight. The fluorine nucleus offers a much greater range of chemical shifts and it has been possible to observe two sets of pentafluorophenyl groups in the system pentafluorophenylmagnesium bromide in ether by their different ^{19}F signals. Equation 4.5 accounts for the variation in signal intensities as the diaryl compound was added to the mixture. An exchange reaction can be observed at 94° [10].

$$2C_6F_5MgBr \rightleftharpoons (C_6F_5)_2Mg + MgBr_2 \qquad\qquad 4.5$$

Witanowski and Roberts [11] have shown that for dineohexylmagnesium, inversion of configuration at the α carbon occurs. The neohexyl group has an interesting NMR spectrum; the α- and β-methylene groups of a configurationally stable derivative give a complex $AA'BB'$ spectrum, but an A_2B_2 spectrum results for configurationally unstable derivatives. The α-methylene protons in dineohexylmagnesium show a complex pattern analysed as the AA' part of an $AA'BB'$ spectrum at 60 MHz in ether solution at 30°C. This collapses at higher temperatures to the triplet expected for the A_2 part of an A_2B_2 system. If the averaging of the coupling constants causing the collapse to A_2B_2 results from internal rotation around the C—Mg bond, then the sum $J_{AB} + J_{AB}'$ should decrease significantly since it is dependent on the population of rotational isomers. The sum $J_{AB} + J_{AB}'$ remains constant throughout. An alternative explanation to internal rotation is inversion of configuration at the α-carbon. This is consistent with all of the NMR evidence. Kinetic data point to this inversion occurring through an S_E1 reaction (via a carbanion). The rate is independent of concentration and has an activation energy of 83.7 ± 8.4 kJ mole^{-1}, and an A factor of $10^{13 \pm 1.5}$ sec^{-1}.

Whitesides and Roberts have used similar means to show that for ether solutions of Grignard reagents, RCH_2CH_2MgX, there is rapid inversion of the α-methylene carbon, possibly through a rate step of order greater than one in Grignard concentration [12]. The NMR spectra of the methylene protons are triplets at room temperature, but on cooling display a more complex pattern. The averaging of the vicinal coupling between α- and β-methylene protons occurs at much lower temperatures than those for the dialkylmagnesium discussed in the preceding paragraph.

Early radiotracer experiments in Grignard systems with radioactive magnesium are now open to some doubt [13, 14] since repeated experiments have not given consistent results with different grades of magnesium. It is now thought that a statistical exchange of magnesium occurs between diethylmagnesium and magnesium bromide [13, 14].

III. Zinc

Calingaert was unable to separate methylethylzinc from the aluminum chloride-catalyzed reaction of dimethyl- and diethylzinc [15] by simple distillation, but the rate of exchange of dimethylzinc with other Group II alkyls [16] is such as to suggest a labile random system was probably present. Mixtures of individual alkyls (Et, Me, and t-Bu) showed no evidence of scrambling in NMR spectra in nonpolar solvents. However, only one set of ethyl absorptions (methyl triplet and methylene quartet) appeared in diethylzinc–ethylzinc halide mixtures. Here the chemical shift varied linearly with the mole fraction of ethylzinc halide, indicating exchange with a preexchange lifetime for alkyl groups on zinc of <0.004 sec [17]. A novel redistribution reaction occurs in the dimethylzinc–methylzinc methoxide system 4.6 and Eisenhuth and Van Wazer have extracted thermodynamic and kinetic data from NMR studies of hydrogen nuclei present [18]. Since

$$2Me_4Zn_4(OMe)_4 \underset{I}{\overset{K}{\rightleftharpoons}} ZnMe_2 + \underset{II}{Me_6Zn_7(OMe)_8} \qquad 4.6$$

dimethylzinc is monomeric but the methoxide (I) is tetrameric with the bird-cage structure [19], the redistribution has a most unusual stoichiometry. The structure of compound I is that of a cube with alternate Zn and O atoms, while that of compound II probably consists of two such cubes joined at a zinc corner. The random equilibrium constant for such a system would be 4, but the observed constant K is 0.068 at 20°, although the enthalpy of reaction is less than 4 kJ mole^{-1}. There is thus a considerable entropy change, -21.5 J °K^{-1}. The spectra also suggest that zinc-bonded methyl groups exchange in a second-order process with an activation energy of about 29 kJ mole^{-1} [18]. This and closely related systems have been studied by several groups of workers [18, 20–22]; they are normally generated by the reaction of alcohol on dialkylzinc in hydrocarbon solvents (4.7). Alkane

$$MeOH + MeZnMe \rightarrow MeH + (1/n)(MeZnOMe)_n \qquad 4.7$$

is evolved and the "bird cage" alkoxide (I) generated disproportionates to the double bird cage (II) for the reaction of methanol with methyl- or phenylzinc, and for ethanol, but not isopropanol nor t-butanol with methylzinc.

The tetrameric alkylzinc alkoxides will themselves redistribute, the example in reaction 4.8 requiring about 20 min at 80° in solution [20] but

$$(MeZnOMe)_4 + (MeZnO\text{-}t\text{-}Bu)_4 \rightarrow Me_4Zn(OMe)_n(O\text{-}t\text{-}Bu)_{4-n} \qquad 4.8$$

the mixture is much more readily generated by reaction of equimolar amounts of ethanol and t-butanol with dimethylzinc. Nuclear magnetic resonance studies of reaction 4.8 indicate four separate signals for methyl attached to zinc, at -0.35, -0.30, -0.23, and -0.14 ppm. The tetraethoxide and the tetrabutoxide absorb, respectively, at -0.35 and -0.14 ppm. The number of signals is best accounted for on the "local environment" approximation, since the local environment of each methyl attached to zinc is composed of three bridging alkoxide groups (see structure I). For a scrambling reaction like 4.8, four such local environments are possible (see also Chapter 3): III, IV, V, and VI. Intramolecular switching of bridging atoms

OR	OR′	OR′	OR′
Zn··Me	Zn··Me	Zn··Me	Zn··Me
RO OR	RO OR	RO OR′	R′O OR′
III	IV	V	VI

in the alkoxides would permit five different environments and hence five methylzinc signals to be observed.

Exchange in the system Me_2Zn and MeZnOMe is more rapid in pyridine than in toluene and a comparison of rate indicates the following decreasing order for alkoxide MeO > EtO > i-PrO > t-BuO [20].

Nuclear magnetic resonance has played a key role in the examination of these zinc systems and in the identification of the new double-birdcage structure. Possibly, many unusual compounds may be found in related systems and it is clear that the exchange processes possible with such compounds will be legion. A large family of compounds can be obtained by reaction of alcohols, phenols, thiols, etc., with dialkylzinc. The structures of the derived RZnXR oligomers are variable, and it is clear that considerable scope exists for the study of exchange reactions. For example, the crystalline methylzinc-t-butyl sulfide is pentameric with a zinc skeleton approximating to a square-based pyramid, while the corresponding isopropyl sulfide is octameric in the crystal with each zinc bonded to three sulfur atoms—compare structures I and II [23, 24]. Polyfunctional acids such as 1,2-, 1,3-, 1,4-, etc. diols, dithiols, amino alcohols, and β-diketones introduce further modifications, e.g., dimeric and trimeric aggregates [25]. Therefore, considerable ingenuity may be needed to decipher the resultant equilibria, especially in solution, even with a guide from crystallography.

Exchange of zinc between zinc acetylacetonate and zinc oxinate in dioxane solution was followed via ^{65}Zn radiotracer, being complete in less than 30 sec, but a heterogeneous system with solid oxinate reacted much more slowly [26].

The distribution ratios of mixed zinc halides (Cl, Br, I) containing ^{65}Zn tracer were examined by Zangen [27] between the two phases—a lithium nitrate–potassium nitrate eutectic mixture of mole ratio 3:4, and an organic eutectic mixture of biphenyl and o- and m-terphenyls in the mole ratio 37:48:15. The equilibrium constant K_{XY} for reaction 4.9 was obtained at

$$ZnX_2 + ZnY_2 \rightleftharpoons 2ZnXY \qquad\qquad 4.9$$

several temperatures for each phase, and the values are given in Table 4.3. There is little temperature change for K in the aromatic solvent but a very significant change in the molten salt phase. Many similar comparisons of

TABLE 4.3

EQUILIBRIUM CONSTANTS K_{XY} FOR THE REACTION $ZnX_2 + ZnY_2 \rightleftharpoons 2ZnXY$[a]

Medium	ClBr	ClI	BrI
LiNO$_3$–KNO$_3$	-1.04 ± 0.2[b]	-1.18 ± 0.2[b]	-1.28 ± 0.2[b]
Polyphenyl	0.93 ± 0.3[b]	0.78 ± 0.3[b]	0.59 ± 0.3[b]
LiNO$_3$–KNO$_3$	1.30 ± 0.1[c]	1.15 ± 0.1[c]	0.98 ± 0.1[c]
Polyphenyl	0.93 ± 0.1[c]	0.75 ± 0.1[c]	0.61 ± 0.1[c]

[a] From Zangen [27].
[b] $\log K_{XY}$ at 150°.
[c] $\log K_{XY}$ at 200°.

equilibria in molten salt, polyphenyl, and aqueous phases have been made, particularly for mercury complexes, and the area has recently been reviewed by Eliezer and Marcus [28]. Equilibria between the anionic mixed halides of zinc (Eq. 4.10) were also studied in the molten salt phase by the same distribution technique. Addition of a large excess of the appropriate halide ion to a zinc halide caused the anionic complexes to appear in the molten salt phase, while only molecular halides were present in the organic phase.

$$ZnX_{i+1}Y_{j-1} + ZnX_{i-1}Y_{j+1} \xrightleftharpoons{K_{ij}} 2ZnX_iY_j \qquad 4.10$$

The constants are given in Table 4.4 with the values for a purely statistical distribution of halogen for comparison. Considerable change with temperature is apparent. In general, the mixed complexes are less stable than expected for a statistical distribution except for the more symmetrical C_{2v} system $ZnCl_2Br_2^{2-}$ [29].

TABLE 4.4

EQUILIBRIUM CONSTANTS K_{ij} FOR THE REACTION OF ANIONIC ZINC COMPLEXES[a, b]

X	Y	Temp (°C)	log K_{11}	log K_{21}	log K_{12}	log K_{31}	log K_{22}
Cl	Br	150	−1.08	−0.26	0.48	−0.68	—
		200	1.28	−0.15	0.58	−0.68	1.30
Cl	I	150	−1.22	−0.29	—	—	—
		200	1.11	−0.16	—	−0.38	—
Br	I	150	−1.35	—	—	—	—
		200	0.96	—	—	—	—
Statistical log K			0.60	0.48	0.48	0.43	0.35

[a] From Zangen [29].
[b] $ZnX_{i+1}Y_{j-1} + ZnX_{i-1}Y_{j+1} \rightleftharpoons 2ZnX_iY_j$ in molten (Li, K)NO$_3$.

IV. Cadmium

We may expect cadmium compounds to participate in redistribution reactions to parallel those of zinc. Simple alkyl self-exchange is solvent catalyzed [30]. Observation of the satellites in proton spectra due to coupling of alkyl hydrogen to ^{111}Cd and ^{113}Cd is a considerable asset in the interpretation of the exchange. The satellites in dimethylcadmium spectra disappear in solvents such as tetrahydrofuran and pyridine and in the presence of methylcadmium methoxide [16, 30]. Although methylcadmium methoxide appears to be isostructural with its zinc counterpart, no disproportionation corresponding to Eq. 4.6 has yet been observed. The preexchange lifetime τ of methyl groups of cadmium has been measured for toluene solutions at 120°, in the presence of added methanol (which forms methylcadmium methoxide); data are shown in Table 4.5. Mixed molecular halides of

TABLE 4.5

Lifetime of Methyl Groups on Me_2Cd in
Toluene at 120°C in the Presence of Methanol[a]

Me_2Cd (moles liter^{-1})	MeOH (moles liter^{-1})	τ (sec)
0.45	0	0.23
0.58	0.0040	0.014
0.56	0.010	0.011
0.48	0.020	0.0072

[a] Activation energy for exchange = 66.8 kJ mole^{-1} [30].

cadmium have been observed in the lithium–potassium nitrate eutectic mixture by Zangen (compare the zinc halides). They are formed in statistical amounts at temperatures of 150–200° [28, 31]. Anionic halocadmates are known [32] from the Raman spectra of the mixed tetrahalocadmates (Br and I) in aqueous solution. The mixed halides were identified through the change in the sharpest line of the T_d species, the symmetrical breathing mode, as the Br/I ratio was altered. Equilibria in perchlorate salt solutions have been examined polarographically [33] and potentiometrically [34] by other workers and the results are quoted in Table 4.6.

TABLE 4.6

Equilibrium Constants K_{ij} for the Reaction $(i/n)CdX_n + (j/n)CdY_n \rightleftharpoons CdX_iY_j$
$(n = i + j)$ Given as $\log K_{ij}$

X_i	Y_j	Medium	Temp (°C)	$\log K$	Method	Ref.
Cl	Br	Nitrate	150–200	0.7	Distribution	31
Cl	I	Nitrate	150–200	0.7	Distribution	31
Br	I	Nitrate	150–200	0.7	Distribution	31
Br	I	2 M NaClO$_4$	25	0.49	Polarographic	33
Cl$_2$	Br	Halide	25	0.14	Potentiometric	34
Cl	Br$_2$	Halide	25	0.48	Potentiometric	34
Cl$_2$	I	Halide	25	0.59	Potentiometric	34
Cl$_2$	I$_2$	6 M NaClO$_4$	25	0.59	Potentiometric	34
Cl	I$_3$	6 M NaClO$_4$	25	0.20	Potentiometric	34

V. Mercury

Exchange reactions involving mercury alkyls have been extensively studied. Reactions of two mercury alkyls (Eq. 4.11) are usually very slow.

$$R_2Hg + R_2'Hg \rightleftharpoons 2RR'Hg \qquad\qquad 4.11$$

However, Calingaert found that methyl and ethyl groups exchanged at room temperature [15] on mercury when aluminum chloride catalyst was used. Without benefit of a catalyst, more exacting temperatures were required. Dimethyl- and perdeuteriodimethylmercury after 78 hr at 65° had exchanged alkyl groups [35]. Calingaert's experiment with a catalyst has been reinvestigated [36] by NMR. Dimethylmercury in benzene at room temperatures has a simple proton spectrum with a central peak (intensity 5) and satellites (combined intensity 1) from coupling to the 16.9 % abundant ^{199}Hg isotope. In the presence of 10 % aluminum chloride there is no ^{1}H–^{199}Hg doublet, and the lifetime of the methyl group before exchange is <0.003 sec.

Quantitative equilibrium studies are recorded in Table 4.7 for Eq. 4.11. These are random for simple alkyl groups but for perfluoro-substituted groups, wide deviations from random are observed [37, 38].

TABLE 4.7

EQUILIBRIUM CONSTANTS $K_{eq} = [RR'Hg]^2/[R_2Hg][R_2'Hg]$ FOR EQ. 4.11[a]

R	R'	K_{eq}	Temp (°C)	Method	Ref.
Me	Et	1.8	90	GLC	37
Et	cyclo-Pr	130	90	GLC	37
Et	Vinyl	86	90	GLC	37
Et	Ph	5.0	90	NMR	37
Vinyl	Ph	15	90	NMR	37
Vinyl	cyclo-Pr	No reaction	90	GLC	37
Et	n-Pr	4.5	90	GLC	37
Et	i-Pr	5.3	90	GLC	37
n-Pr	i-C$_3$F$_7$	2×10^{-3}	90	GLC	37
i-Pr[b]	n-C$_3$F$_7$	2×10^{-3}	90	GLC	37
Me[c]	Et	3	25	Distillation	15
Me[d]	Ph	4.4	150	NMR	38

[a] $K_{random} = 4$. Reagents neat unless otherwise noted.
[b] Toluene solvent.
[c] Catalyst Al$_2$Cl$_6$.
[d] Time required, 6 hr.

Mixed halides and pseudohalides of mercury are well documented [28, 39] and equilibrium data from an assortment of sources are given in Table 4.8. For the molecular halides, it is possible to compare equilibrium constants in widely different media. Results at 25° in aqueous solutions and in nonpolar solvent benzene are remarkably similar for simple Cl, Br, or I mixed halides and quite close to random ($\log K_{\text{random}} = 0.6$). At high

TABLE 4.8

EQUILIBRIUM CONSTANTS FOR THE FORMATION OF MIXED HALIDES AND PSEUDOHALIDES
OF MERCURY $K_{XY}{}^a$

X	Y	$\log K_{XY}$	Ref.	X	Y	$\log K_{XY}$	Ref.
Fused Li–KNO$_3$ at 150°C				Methanol at 25°C (*continued*)			
Cl	Br	1.88	40	I	SCF$_3$	0.22	44
Cl	I	−0.34	40	SCN	SCF$_3$	0.13	44
Br	I	−0.94	40	Cl	SeCF$_3$	1.30	44
				Br	SeCF$_3$	1.18	44
Water at 25°C				I	SeCF$_3$	0.31	44
Cl	CN	(0.92)	41	SCN	SeCF$_3$	0.37	44
Br	CN	(0.3)	41				
I	CN	(−0.24)	41				
				Polyphenyl at 150°C			
0.5 M NaClO$_4$ at 25°C				Cl	Br	1.69	45
Cl	Br	1.20	42, 43	Cl	I	1.65	45
Cl	I	1.75	42, 43	Br	I	1.58	45
Br	I	1.10	42, 43				
				Benzene at 25°C			
Methanol at 25°C				Cl	Br	1.16	46
Cl	SCF$_3$	1.30	44	Cl	I	1.50	46
Br	SCF$_3$	0.36	44	Br	I	0.76	46

^{a} $\text{HgX}_2 + \text{HgY}_2 \xrightarrow{K_{XY}} 2\text{HgXY}$, $K_{\text{random}} = 4$.

temperatures (150°) in molten salts the constants are more variable, but in the nonpolar polyphenyl melt are similar to those at 25°. Data for the anionic mixed halides $\text{HgX}_i\text{Y}_j^{m-}$ have been obtained in fused salts (by distribution methods using the γ radiation of ^{197}Hg to analyze mercury concentration of each phase) at one temperature only, and show lack of stability of mixed iodides in particular [28, 40]. However, recent work with zinc [29] in molten salts has indicated considerable variation with temperature in corresponding equilibria for anionic zinc halides, and it may be premature to comment at this state on the results at a single temperature [40].

There have been many mechanistic studies of alkyl exchange between the various mercury alkyl species and mercury halides and alkylmercuric halides [47–55], which are shown in Eqs. 4.12–4.14. Equation 4.12 represents

$$R_2Hg + R'Hg^*X \rightleftharpoons R'RHg^* + RHgX \qquad 4.12$$

$$R_2Hg + HgX_2 \rightleftharpoons 2RHgX \qquad 4.13$$

$$RHgX + Hg^*X_2 \rightleftharpoons RHg^*X + HgX_2 \qquad 4.14$$

the slowest of these alkyl exchanges. It has been examined kinetically by a double-labeling technique, the alkyl group being optically active (e.g., *sec*-butyl) and the mercury labeled with a radioactive tracer (^{203}Hg) [49]. The kinetics are second order overall, first order in each reagent, and the configuration of the optically active carbon attached to mercury is unchanged in the reaction [49]. One alkyl group is exchanged per mercury transfer. It becomes clear from these results that the mechanism is of electrophilic attack, mercury attacking carbon *on the same side*† as the leaving group (RHg$^+$). This might be by the S_F2 (four-center) transition state (VII) or the S_E2 (open) transition state (VIII). The rate is greater for

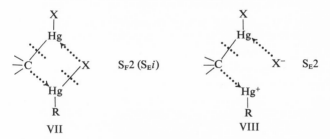

anions X with the lowest affinity for mercury (X = NO$_3$ > OAc > Br) (see Table 4.9). This is consistent with an S_E2 mechanism, since the ionization of the group X provides RHg$^+$ which should be a better electrophile than the neutral RHgX molecule, and in addition the X$^-$ is unlikely to close the four-center transition state (VIII) if it has a low affinity for mercury.

Reaction 4.13 favors the mixed compound exclusively. Heats of redistribution were measured calorimetrically [56] where R is Me, Et and X is Cl,

---

† The usual mechanism for bimolecular nucleophilic substitution on carbon S$_N$2 is, of course, backside attack with attacking and leaving *nucleophiles* on *opposite* sides of the carbon.

TABLE 4.9

RATE CONSTANTS IN THE SYSTEMS RHgX + HgX_2 OR HgR_2, ILLUSTRATING SALT EFFECTS[a] FOR EQUIMOLAR REAGENTS IN ETHANOL

RHgX	R_2Hg or HgX_2	Temp (°C)	k_2 (liter $mole^{-1}$ sec^{-1})
sec-BuHgBr	$(sec\text{-}Bu)_2Hg$	35	4.6×10^{-5} (normal salt effect)
sec-BuHgOAc	$(sec\text{-}Bu)_2Hg$	35	2.7×10^{-4}
sec-$BuHgNO_3$	$(sec\text{-}Bu)_2Hg$	0	3.4×10^{-2} (normal salt effect)
MeHgBr	$HgBr_2$	100.2	1.28×10^{-4}
MeHgI	HgI_2	100.2	10.1×10^{-4}
MeHgOAc	$Hg(OAc)_2$	59.8	5.0×10^{-3}
$MeHgNO_3$	$Hg(NO_3)_2$	0	16.9×10^{-4}
MeHgBr	$HgBr_2$	59.8	0.49×10^{-5}
MeHgBr[b]	$HgBr_2$	59.8	39×10^{-5} (specific positive
MeHgBr[c]	$HgBr_2$	59.8	88×10^{-5} salt effect)

[a] From Charman *et al.* [49, 50] and Hughes *et al.* [51].
[b] Contains LiBr in mole ratio to $HgBr_2$ of 1.1.
[c] Contains LiBr in mole ratio to $HgBr_2$ of 2.14.

Br, or I (Table 4.10) and confirmation of their magnitude comes from more recent NMR work. The neat liquids dimethylmercury and mercuric halide were scrambled at 150° and later analyzed at room temperature in deuteriochloroform by ^1H NMR [38].

Many kinetic investigations have been made in this system. Rausch and Van Wazer [38] most recently confirmed second-order kinetics where R is Me, Et and X is Cl, Br, and I in dioxane and methanol, by measuring change in signal integral with time, and obtained rate constants which compared

TABLE 4.10

ENTHALPIES OF THE SCRAMBLING REACTION $R_2Hg + HgX_2 \rightleftharpoons 2RHgX$[a, b]

R	X	ΔH (kJ $mole^{-1}$)	R	X	ΔH (kJ $mole^{-1}$)
Me	Cl	27.6	Et	Cl	30.6
Me	Br	19.3	Et	Br	21.4
Me	I	13.4	Et	I	10.9

[a] From Skinner [56].
[b] Data refer to gas phase.

satisfactorily with those obtained using UV spectroscopy at lower concentrations [52, 53]. Where R is p-XC$_6$H$_4$—, there is a Hammet $\sigma\rho$ correlation of rate constants k with a high negative ρ (−5.87), showing that the point of electrophilic attack is the carbon attached to mercury. Rate constants in general increase with ionicity of mercuric salt (Cl > Br > I) in keeping with an ion pair or S_E2-type mechanism. The variation of rate parameters with substituent and solvent for these second-order reactions is shown in Table 4.11. Ingold and co-workers found second-order kinetics and retention of

TABLE 4.11

RATE CONSTANTS AND ACTIVATION ENERGIES FOR THE REACTION R$_2$Hg + HgI$_2$ $\rightleftharpoons$ 2RHgI
AT 25°C

Solvent	R	k (liter mole^{-1} sec^{-1})	$E^{\ddagger}$ (kJ mole^{-1})	$\Delta S^{\ddagger}$ (J °K^{-1})	Ref.
Dioxane	Et	1.63×10^{-2}	51.4	−117	52
Dioxane	n-Pr	1.86×10^{-2}	51.1	−117	52
Dioxane	i-Pr	1.60×10^{-2}	50.2	−121	52
Dioxane	cyclo-Pr	7.67×10^{-2}	53.6	−92	52
Dioxane	p-ClC$_6$H$_4$	9.2×10^{-2}	60.6	−71	52
Dioxane	CF$_3$	Very slow	—	—	38
Ethanol	Ph	62.8	48.9	−54	53
Benzene	Ph	29.2	31.8	−117	53
Cyclohexane	Ph	15.9	31.8	−130	53
Dioxane	Ph	1.97	53.6	−67	52

configuration in this system also (compare Eq. 4.12) and additional evidence for an S_E2 transition state in a specific salt effect. A specific retarding salt effect of bromide on mercuric bromide exchange was ascribed to the pre-equilibrium which could block the formation of an S_E2 transition state,

$$HgBr_2 + Br^- \rightleftharpoons HgBr_3^-$$

since HgBr$_3^-$ would be much less electrophilic than HgBr$_2$ [48].

The reverse of Eq. 4.13, the so-called symmetrization reaction, can be conducted if a strong amine is added to remove the HgBr$_2$ from the system and so force the symmetrization to occur [54, 55]. Reutov and co-workers [57] have shown that this is an electrophilic displacement, second order in RHgX for the compound where R is p-YC$_6$H$_4$CH(CO$_2$Z), X is Br or I, and Z is alkyl. The rate is affected by substituents as follows: Z = Me > Et > i-Pr > t-Bu and Y = Cl > Br > H > Me. Reaction of the ester

PhCH(HgBr)CO$_2$Et with labeled phenylmercuric bromide in pyridine is also second order (possibly S$_E$2) [58]. Another variant on this reaction is the exchange between substituted benzylmercuric bromides which is always S$_E$2 [55].

For the one-alkyl exchange reaction shown in Eq. 4.14, Ingold and co-workers endorse the S$_F$2 (VII) rather than the open S$_E$2 (VIII) transition state, again on the basis of a salt effect (Table 4.9). They observed second-order kinetics in this system, first order both in mercury dihalide and in alkylmercuric halide, and retention of configuration of optically active carbon attached to mercury. The telling feature of this investigation, however, was the specific salt effect of halide ions. Whereas these had a minor salt effect on the three-alkyl exchange 4.12, and, mole for mole, a very strong decelerating effect on the two-alkyl exchange 4.13, an extraordinary accelerating effect was observed on the one-alkyl exchange. Since the action of halide must be the same—formation of HgX$_3$$^-$—in the latter cases, it can be stated that the transition state in 4.14 is stabilized by anionic HgBr$_3$$^-$ in contrast to exchange 4.13, where it is destabilized. This is accommodated well by the hypothesis of S$_F$2 transition state for the one-alkyl and S$_E$2 transition for the two-alkyl exchange [50, 51].

The one-alkyl exchange is usually second order [50, 51]; but has been found to be first order (S$_E$1) in dimethyl sulfoxide or 70% aqueous dioxane for the ester PhCH(HgBr)CO$_2$Et [55, 59, 60] and also for p-nitrobenzyl-mercuric bromide in dimethyl sulfoxide [55] in their exchanges with mercuric bromide.

Kinetic effects of substituents have been studied for many examples of reactions 4.12–4.14. In Dessy's work with mercury diaryls the electron-donating groups para to the mercury substituent accelerated reaction as expected when the carbon attached to mercury was the site of electrophilic attack (the S$_E$2 prediction). A similar effect was observed by the Russian workers [61] for S$_E$2 reactions of benzylmercuric bromide. For the S$_E$1 reaction where C–Hg fission in the rate-determining step gives HgBr$^+$, electron-withdrawing substituents should accelerate reaction, as has been observed [55, 59, 60].

Activation energies and entropies can be compared for reactions of the S$_E$1 and S$_E$2 types (Table 4.12); note that positive entropies are found for the S$_E$1 reaction (increase in number of species present):

$$\text{RHgX} \quad \xrightarrow[\text{slow}]{\text{S}_\text{E}1} \quad \text{R}^- + \text{HgX}^+$$

Negative entropies, however, are found for the bimolecular S$_E$2 reaction.

TABLE 4.12

Proposed Mechanism and Activation Parameters for Some Mercury Exchange Reactions Examined by Reutov and Co-workers

Reaction	Mechanism	Solvent	$E^{\ddagger}$ (kJ mole^{-1})	$\Delta S^{\ddagger}$ (J °K^{-1})	Ref.
$R = YC_6H_4CH(CO_2R)$					
$2RHgX \xrightarrow{NH_3} R_2Hg + HgX_2$	S_E2	$CHCl_3$	—	—	57
$RHgBr + PhHgBr \rightleftarrows RHgBr + PhHgBr$	S_E2	Pyridine	50.2	—	58
$RHgBr + HgBr_2 \rightleftarrows RHgBr + HgBr_2$	S_E2	Pyridine	68.2	−65.0	59, 60
$RHgBr + HgBr_2 \rightleftarrows RHgBr + HgBr_2$	S_E1	70% dioxane	112	+19.3	59, 60
$RHgBr + HgBr_2 \rightleftarrows RHgBr + HgBr_2$	S_E1	Me_2SO	—	—	55
$R = p\text{-}YC_6H_4CH_2$					
$RHgBr + HgBr_2 \rightleftarrows RHgBr + HgBr_2$	S_E2	Me_2SO	—	—	55
$RHgBr + HgBr_2 \rightleftarrows RHgBr + HgBr_2$ (Y = H)	S_E2	Quinoline	78.6	—	61
$RHgBr + HgBr_2 \rightleftarrows RHgBr + HgBr_2$ (Y = NO_2)	S_E1	Me_2SO	—	—	55
$R = 1,3\text{-dimethylbutyl}$					
$R_2Hg + RHgBr$	S_E2 (S_{Ei})	Ethanol	64	−13.2	62
$PhHgCl + Hg^*Cl_2$	S_E2 (S_{Ei})	Toluene	25	—	63

Some confusion arose in NMR work on alkylmercuric halides. Sharp satellites (from ^{1}H–^{199}Hg coupling) arise in the ^{1}H NMR spectra of MeMgX, where X is Cl, CN, OAc; but where X is I, SCN, Br, the lines are broadened or completely merged with the central line [64]. It is now clear that broadening in the iodide spectrum is due to rapid relaxation of the iodine quadrupole [36, 65, 66] and that no alkyl–mercury fission occurs on the time scale represented by chemical shifts in this system. Mercury–halogen exchange does, however, occur:

$$MeHgCN + Me^*HgX \rightleftharpoons MeHgX + Me^*HgCN$$

Methylmercuric cyanide and methylmercuric halides exchange halogen and cyanide in dimethylformamide in a second-order reaction.

The thermodynamic parameters are $E^{\ddagger} = 92.4$, 63, and 37.8 kJ mole^{-1} and $\Delta S^{\ddagger} = +16.7$, -75, and -104.6 J °K^{-1} for I, Cl, and Br, respectively. Mercuric halides and Me$_4$N$^+$X$^-$ are catalysts, as are basic solvents [65].

VI.   Exchange of Mixtures of Group II and Other Elements

Certain exchange reactions of Group II metal alkyls with lithium alkyls are mentioned in Chapter 3, Section I. The self-exchange of Group II alkyls is slow (see Sections I–V) but preexchange lifetimes show that mutual exchange of alkyl groups between Zn–Mg, and Cd–Mg alkyls is rapid. An apparent order of approximately 2 was observed [67] for a series of solvents: there was little solvent effect, so it is unlikely the transition state is highly polar. It is also unlikely to go by a free radical mechanism since prohibitive activation energies are required to dissociate the alkyls

$$MMe_2 \rightarrow Me \cdot + MMe \cdot$$

(For Zn, Cd, and Hg, $E^{\ddagger} = 197.5$, 192, 210 kJ mole^{-1}, respectively.)

The exchange of dimethylcadmium with trimethylgallane, trimethylindane, and dimethylzinc has been examined thoroughly by NMR methods [68]. All kinetic parameters were obtained by the complete line-shape analysis of the NMR spectra. The negative entropies and second-order rates are consistent with a four-center mechanism (Table 4.13).

Several studies have appeared in which mercuric halides substitute on metal alkyls (Eqs. 4.15 and 4.16). The kinetics of substitution of ethylzinc by phenylmercuric chloride (Eq. 4.15) are first order in each reagent and

TABLE 4.13

Exchange of Methyl between Me$_2$Cd and Metal Methyl[a]

Metal methyl	$E^{\ddagger}$ (kJ mole^{-1})	$\Delta S^{\ddagger}$ (J °K^{-1})[b]	k (liter mole^{-1} sec^{-1})
GaMe$_3$	32.7 ± 3.4	-79.6	1860
InMe$_3$	35.2 ± 0.8	-75.5	1300
ZnMe$_2$	71.2 ± 4.2	-22.2	0.42

[a] From Henold et al. [68].
[b] Temperature = 25°C.

solvent dependent. The relative rates in ether at 35° for the R$_2$Zn series are R = Me (100), R = Et (450), R = Pr (1700) and R = i-Pr (2200) and an S$_E i$ mechanism is proposed (VII) [69]. The reaction of tetraethyltin with

$$ZnEt_2 + PhHgCl \xrightarrow{\text{S}_{\text{E}}i} PhHgEt + EtZnCl \qquad 4.15$$

$$SnEt_4 + HgCl_2 \xrightarrow{\text{S}_{\text{E}}2} EtHgCl + Et_3SnCl \qquad 4.16$$

mercuric chloride (Eq. 4.16) is also bimolecular, and a thorough investigation of solvation effects on reagents, products, and the transition state has been carried out [70]. The general conclusion was that an open S$_E$2 transition state is involved (VIII).

REFERENCES

1. N. A. Matwiyoff and W. G. Movius, *J. Amer. Chem. Soc.* **89**, 6077 (1967).
2. R. E. Dessy, *J. Amer. Chem. Soc.* **82**, 1580 (1960).
3. E. C. Ashby, R. Sanders, and J. Carter, *Chem. Commun.* **1967**, 997.
4. E. C. Ashby, *Quart. Rev. Chem. Soc.* **21**, 259 (1967).
5. B. J. Wakefield, *Organometal. Chem. Rev.* **1**, 131 (1966).
6. D. Bryce-Smith and G. F. Cox, *J. Chem. Soc.* **1961**, 1175.
7. G. D. Stucky and R. E. Rundle, *J. Amer. Chem. Soc.* **85**, 1002 (1963).
8. M. B. Smith and W. E. Becker, *Tetrahedron* **22**, 3027 (1966).
9. M. B. Smith and W. E. Becker, *Tetrahedron* **23**, 4215 (1967).
10. D. F. Evans and M. S. Khan, *Chem. Commun.* **1966**, 67.
11. M. Witanowsky and J. D. Roberts, *J. Amer. Chem. Soc.* **88**, 737 (1966).
12. G. Whitesides and J. D. Roberts, *J. Amer. Chem. Soc.* **87**, 4878 (1965).

13. R. E. Dessy, S. Green, and R. M. Ṣalinger, *Tetrahedron Lett.* **1964**, 1369.
14. D. O. Cowan, J. Hsu, and J. D. Roberts, *J. Organometal. Chem.* **29**, 3688 (1964).
15. G. Calingaert and H. A. Beatty, *in* "Organic Chemistry. An Advanced Treatise," p. 1806. Wiley, New York, 1950.
16. R. E. Dessy, G. R. Coe, and R. M. Salinger, *J. Amer. Chem. Soc.* **85**, 1191 (1963).
17. J. Boersma and J. G. Noltes, *J. Organometal. Chem.* **8**, 551 (1968).
18. W. Eisenhuth and J. R. Van Wazer, *J. Amer. Chem. Soc.* **90**, 5397 (1968).
19. H. M. M. Shearer and C. B. Spencer, *Chem. Commun.* **1966**, 194.
20. E. A. Jeffery and T. Mole, *Aust. J. Chem.* **21**, 1187 (1968).
21. G. Allen, J. M. Bruce, D. W. Farren, and F. G. Hutchinson, *J. Chem. Soc. B* **1966**, 799.
22. J. M. Bruce, B. C. Cutsforth, D. W. Farren, F. G. Hutchinson, F. M. Rabagliati, and D. R. Reed, *J. Chem. Soc. B* **1966**, 1020.
23. G. W. Adamson and H. M. M. Shearer, *J. Chem. Soc. D* **1969**, 897.
24. G. W. Adamson, H. M. M. Shearer, and C. B. Spencer, *Acta Crystallogr. Suppl.* **21A**, 135 (1966).
25. J. Boersma and J. G. Noltes, *J. Organometal. Chem.* **13**, 291 (1968).
26. K. Saito and M. Tamura, *Bull. Chem. Soc. Japan* **32**, 533 (1959).
27. M. Zangen, *Inorg. Chem.* **7**, 138 (1968).
28. I. Eliezer and Y. Marcus, *Coord. Chem. Rev.* **4**, 273 (1969).
29. M. Zangen, *Inorg. Chem.* **7**, 1202 (1968).
30. N. S. Ham, E. A. Jeffery, T. Mole, J. K. Saunders, and S. N. Stuart, *J. Organometal. Chem.* **8**, P7 (1967).
31. M. Zangen, quoted in Eliezer and Marcus [28].
32. J. Rolfe, D. E. Sheppard, and L. A. Woodward, *Trans. Faraday Soc.* **50**, 1275 (1954).
33. A. Swinarski and A. Grodzicki, *Rocz. Chem.* **41**, 1205 (1967); *Chem. Abstr.* **68**, 108525s (1968).
34. Ya. D. Fridman, D. S. Zarbaev, and R. I. Sorochan, *Russ. J. Inorg. Chem. (English Transl.)* **5**, 381 (1960).
35. R. E. Dessy, F. Kaplan, G. R. Coe, and R. M. Salinger, *J. Amer. Chem. Soc.* **85**, 1191 (1963).
36. N. S. Ham, E. A. Jeffery, T. Mole, and S. N. Stuart, *Chem. Commun.* **1967**, 254.
37. G. F. Reynolds and S. R. Daniel, *Inorg. Chem.* **6**, 480 (1967).
38. M. D. Rausch and J. R. Van Wazer, *Inorg. Chem.* **3**, 761 (1964).
39. G. B. Deacon, *Rev. Pure Appl. Chem.* **13**, 189 (1961).
40. M. Zangen and Y. Marcus, *Israel J. Chem.* **2**, 155 (1964).
41. M. T. Beck and F. Gaizer, *Magyar Kem. Folyorat* **69**, 555, 559 (1963); *Chem. Abstr.* **60**, 7663h, 7664a (1964).
42. Y. Marcus and I. Eliezer, *J. Phys. Chem.* **66**, 1661 (1962).
43. Y. Marcus, *Acta Chem. Scand.* **11**, 610 (1957).
44. H. J. Clase and E. A. V. Ebsworth, *J. Chem. Soc.* **1965**, 940.
45. M. Zangen, *Israel J. Chem.* **2**, 91 (1964).
46. I. Eliezer, *J. Phys. Chem.* **68**, 2722 (1964).
47. H. B. Charman, E. D. Hughes, and C. K. Ingold, *J. Chem. Soc.* **1959**, 2523.
48. H. B. Charman, E. D. Hughes, and C. K. Ingold, *J. Chem. Soc.* **1959**, 2530.
49. H. B. Charman, E. D. Hughes, and C. K. Ingold, *J. Chem. Soc.* **1961**, 1121.
50. H. B. Charman, E. D. Hughes, C. K. Ingold, and H. C. Volger, *J. Chem. Soc.* **1961**, 1142.

51. E. D. Hughes, C. K. Ingold, F. G. Thorpe, and H. C. Volger, *J. Chem. Soc.* **1961**, 1133.
52. R. E. Dessy and Y. K. Lee, *J. Amer. Chem. Soc.* **82**, 689 (1960).
53. R. E. Dessy, Y. K. Lee, and J. Y. Kim, *J. Amer. Chem. Soc.* **83**, 1163 (1961).
54. O. A. Reutov, *Angew. Chem.* **72**, 198 (1960).
55. O. A. Reutov, *Omagiu Raluca Ripan* **1966**, 481; *Chem. Abstr.* **67**, 53253z (1967).
56. H. A. Skinner, *Rec. Trav. Chim.* **73**, 991 (1954).
57. O. A. Reutov and I. P. Beletskaya, *Dokl. Akad. Nauk SSSR* **131**, 853 (1960); *Chem. Abstr.* **54**, 16424d (1960).
58. O. A. Reutov, H. H. Weng, I. P. Beletskaya, and T. A. Smolina, *Russ. J. Phys. Chem.* **35**, 1197 (1961).
59. O. A. Reutov, V. I. Solokov, and I. P. Beletskaya, *Izv. Akad. Nauk SSSR, Otd. Khim. Nauk* **1961**, 1213, 1217, 1427, 1561; *Chem. Abstr.* **58**, 12386c, e, f (1963).
60. O. A. Reutov, V. I. Solokov, and I. P. Beletskaya, *Dokl. Akad. Nauk* **136**, 631 (1961); *Chem. Abstr.* **55**, 17557d (1961).
61. O. A. Reutov, T. A. Smolina, and V. A. Kalyavin, *Dokl. Akad. Nauk SSSR* **139**, 389 (1961); *Chem. Abstr.* **56**, 1470b (1962).
62. T. P. Karpov, V. A. Malyanov, E. V. Uglova, and O. A. Reutov, *Izv. Akad. Nauk. SSSR Ser. Khim.* **1964**, 1580; *Chem. Abstr.* **62**, 411b (1965).
63. T. A. Smolina, M. Chou, and O. A. Reutov, *Izv. Akad. Nauk SSSR Ser. Khim.* **1966**, 413; *Chem. Abstr.* **64**, 18483e (1966).
64. J. V. Hatton, W. G. Schneider, and W. Siebrand, *J. Chem. Phys.* **39**, 1330 (1963).
65. L. L. Murrell and T. L. Brown, *J. Organometal. Chem.* **13**, 301 (1968).
66. D. N. Ford, P. R. Wells, and P. C. Lauterbur, *Chem. Commun.* **1967**, 616.
67. C. R. McCoy and A. L. Allred, *J. Amer. Chem. Soc.* **84**, 912 (1962).
68. K. Henold, J. Soulati, and J. P. Oliver, *J. Amer. Chem. Soc.* **91**, 3171 (1969).
69. M. H. Abraham and P. H. Rolfe, *J. Organometal. Chem.* **7**, 23 (1967).
70. M. H. Abraham, G. F. Johnston, J. F. S. Oliver, and J. A. Richards, *Chem. Commun.* **1969**, 930.

5

Transition Metals

There has not been much activity in the study of redistribution reactions of transition metals either of the d or f series, except nickel. In this chapter, it will be convenient to group the material under ligands rather than under specific elements; thus, data for any one element may appear in several sections. In this way families such as the eight-coordinated β-diketonates $M(acac)_4^-$ and $M'(acac)_4$, where M is Y or La and M' is Ce, Th, U, Zr, or Hf, can be dealt with together.

I. Carbonyls

Certain instances of redistribution have been noted for transition metal ligands thought to bond synergically, principally CO, PF_3, and other phosphines (Eq. 5.1). Scrambling for reactions of CO alone or with PF_3 on

$$Ni(CO)_4 + Ni(PF_3)_4 \rightarrow Ni(CO)_3PF_3 + Ni(CO)_2(PF_3)_2 + Ni(CO)(PF_3)_3 \quad 5.1$$

nickel have been investigated mainly from the kinetic viewpoint. Random scrambling was semiquantitatively demonstrated for the $Ni(CO)_n(PF_3)_{4-n}$ system [1] using GLC techniques. Perhaps the most fundamental study was of the scrambling of isotopic forms of CO on nickel (Eq. 5.2). Recently the

$$Ni(C^{16}O)_4 + Ni(C^{18}O)_4 \rightarrow Ni(C^{16}O)_{4-n}(C^{18}O)_n \quad 5.2$$

scrambling kinetics in both gas phase and solution have been measured with arresting results [2]. The scrambling reaction is over in 1 min in the gas phase at 25°, but incomplete at the same temperature in hexane after 24 hr! However, there is a concurrent reaction in which free CO substitutes on $Ni(CO)_4$ and the ratio $k_{scrambling}/k_{substitution} \gg 1$ in gas and $\ll 1$ in hexane, so that the mechanism of scrambling in hexane may well be dissociation to CO followed by substitution. The entire system may be subject to heterogeneous catalysis and it is clearly necessary to define conditions carefully when discussing exchange data and when attempting to compare different systems.

Clark and Brimm [1] looked at the scrambling reaction given by Eq. 5.1 at 75° in the gas phase and found it to be approximately random with the more heavily PF_3-substituted complexes preferred. Individual mixed complexes are sufficiently labile to disproportionate on storing in sealed tubes as neat liquids at room temperature (Table 5.1). The more heavily

TABLE 5.1

EXTENT OF DISPROPORTIONATION OF LIQUID COMPLEXES AT ROOM TEMPERATURE[a]

Reaction	Extent
$Mo(PF_3)(CO)_5 \rightleftharpoons Mo(CO)_6 + Mo(PF_3)_2(CO)_4$ (cis–trans)	10%, 60 min
$Ni(CO)_3PF_3 \rightleftharpoons Ni(CO)_4 + Ni(CO)_2(PF_3)_2$, etc.	20%, 10 min
$Ni(CO)_2(PF_3)_2 \rightleftharpoons Ni(CO)_4 + Ni(CO)_3(PF_3)$, etc.	20%, 100 min[b]
$Ni(CO)(PF_3)_3 \rightleftharpoons Ni(CO)_4 + Ni(CO)_2(PF_3)_2$, etc.	20%, 480 min

[a] From Clark and Brimm [1] and Clark and Hoberman [10].
[b] Time > 700 min for CO atmospheres suggests a dissociative mechanism.

PF_3-substituted complexes disproportionate most slowly, and the time required for disproportionation is greatly increased if the complex is in an atmosphere of CO. This latter fact suggests a dissociative mechanism, in which CO dissociates in a preequilibrium (Eq. 5.3) suppressed by excess CO. This could be followed by bimolecular substitution reactions such as Eqs. 5.5 and 5.4 as shown, or further dissociation and recombination reactions

$$Ni(CO)_2(PF_3)_2 \rightleftharpoons Ni(CO)(PF_3)_2 + CO \qquad 5.3$$

$$CO + Ni(CO)_2(PF_3)_2 \rightleftharpoons Ni(CO)_3(PF_3) + PF_3 \qquad 5.4$$

$$PF_3 + Ni(CO)_2(PF_3)_2 \rightleftharpoons Ni(CO)(PF_3)_3 + CO \qquad 5.5$$

which could likewise lead to scrambling. Basolo [3] has recently reviewed work on substitution in metal carbonyls and concludes that mechanisms are essentially dissociative where nucleophiles like CO and PF_3 are concerned. There is undoubtedly a connection with the rate of CO substitution in the complexes for most of the scrambling reactions of CO and PF_3 studied thus far, but comparisons are hampered by the lack of complete kinetic information in different phases for all the reactions concerned. With this firmly in mind, some qualitative differences for different metal systems are now discussed.

The mixed $Fe(CO)_x(PF_3)_{5-x}$ complexes do not readily disproportionate [4, 5]. A rapid exchange between isomers of any one stoichiometry is thought to be due to rapid intramolecular flipping and not to any intermolecular process. The mixed manganese pentacarbonyl hydrides $HMn(CO)_x(PF_3)_{5-x}$ and the related perfluoroalkyl derivatives $R_fMn(CO)_x(PF_3)_{5-x}$ show no tendency to disproportionate via scrambling reactions [6]. The hydrides, although six coordinate, are not stereochemically rigid* and again intramolecular switching of groups is postulated. Some PF_3-substituted derivatives of decacarbonyldimanganese with one, two, and three PF_3 substituents have been made and again the neat complexes show no tendency to scramble [7]. In each of these Fe and Mn systems, the presence of excess ligand gas is necessary for scrambling to occur, which suggests scrambling can only occur via substitution, and not via four-center mechanisms. It is interesting to note that CO exchange with $Fe(CO)_5$, $Mn_2(CO)_{10}$, and $Mn(CO)_5X$ occurs at a very much lower rate than for $Ni(CO)_4$, for which scrambling also is rapid.

In similar experiments with the five-coordinate cobalt systems $RCo(CO)_{4-x}(PF_3)_x$, where R is H, CF_3, C_2F_5, or C_3F_7, the hydrides containing least PF_3 substituents were found to scramble so rapidly that they could not be completely purified by GLC techniques [8]. The rate of CO exchange with $HCo(CO)_4$ is high [9]. Scrambling in the perfluoroalkyl systems is slower. Scrambling in the hydride series departed considerably from random; the presence of Co—H bonds in the system clearly favors bonding to PF_3, which is thought to have the stronger synergic bonding ability.

Scrambling also occurs in the octahedral $Mo(CO)_x(PF_3)_{6-x}$ system, for which all possible mixed compounds, including three sets of cis–trans isomers are known [10]. In Table 5.1 a rough comparison is made with rates of some nickel disproportionations. The rate of disproportionation of the

---

* Cis and trans isomers of related compounds have recently been detected by NMR [6a].

monophosphine–Mo complex was very much less than that of CO or PF_3 substitution of the compound (Eq. 5.6).

$$2Mo(PF_3)(CO)_5 \rightleftharpoons Mo(CO)_6 + Mo(PF_3)_2(CO)_4 \qquad 5.6$$

Another scrambling reaction conveniently included in this section is shown in Eq. 5.7. This is technically scrambling of $Mo(CO)_5$ and $W(CO)_5$

$$\underset{\text{I}}{Mo_2H(CO)_{10}^-} + \underset{\text{II}}{W_2H(CO)_{10}^-} \rightleftharpoons \underset{\text{III}}{2MoWH(CO)_{10}^-} \qquad 5.7$$

moieties on bridging hydrogen as central metal ion! The anions I and II were provided as their tetraethylammonium salts in tetrahydrofuran solutions and were observed to scramble at 80°C over a period of 60 hr to give the mixed anion III in statistical amounts. The reaction was followed by the ^1H NMR signals which lie at τ 22.15 (I), 22.52 (II), and 22.37 (III) and are readily identified through the coupling to ^{183}W, the only isotope of spin $\frac{1}{2}$ (14.28% abundant) present [11].

II. Dithiolates, Salicylaldiminates, and Aminotroponiminates

Exchange of dithiolate ligands on nickel(II) square-planar complexes has been monitored by polarographic techniques (see Chapter 2, Section IV) in methylene chloride or acetonitrile solution (Eq. 5.8). The reaction times

$$\underset{\text{IV}}{[R_2C_2S_2]_2Ni} + \underset{\text{V}}{[R_2'C_2S_2]_2Ni} \rightleftharpoons \underset{\text{VI}}{2[R_2'C_2S_2]Ni[S_2C_2R_2]} \qquad 5.8$$

noted in Table 5.2 were observed for reactions with various substituents R and R′ [12]. The reaction favors the mixed complex in all the reactions studied.

TABLE 5.2

EXTENT AND TIME OF REACTION 5.8 FOR VARIOUS SUBSTITUENTS R AND R′[a] IN IV, V, AND VI

R	R′	Solvent	Time (days)	Temp (°C)	Mixed product (%)
Ph	CF_3	CH_3CN	13	25	92
Ph	CF_3	CH_2Cl_2	6	40	77
Ph	CN	CH_2Cl_2	6	40	—
CF_3	CN	CH_2Cl_2	6	40	—

[a] From Davison et al. [12].

Statistical ligand distribution was found in the equilibration of bis(salicylaldimino)–Ni(II) complexes (VII) containing various substituents in the aromatic rings (X) and on amine nitrogen (R) [Eqs. 5.9–5.11] (13).

VII

Structure VII is represented in the equations by Ni[R(sal)X][R'(sal)X'].

$$Ni[t\text{-Bu(sal)}H]_2 + Ni[i\text{-Pr(sal)}H]_2 \rightleftharpoons 2Ni[t\text{-Bu(sal)}H][i\text{-Pr(sal)}H] \qquad 5.9$$

$$Ni[t\text{-Bu(sal)}Me]_2 + Ni[t\text{-Bu(sal)}H]_2 \rightleftharpoons 2Ni[t\text{-Bu(sal)}Me][t\text{-Bu(sal)}H] \qquad 5.10$$

$$Ni[t\text{-Bu(sal)}Me]_2 + Ni[i\text{-Pr(sal)}H]_2 \rightleftharpoons 2Ni[t\text{-Bu(sal)}Me][i\text{-Pr(sal)}H] \qquad 5.11$$

The ligands apparently remain intact during the scrambling process, for the complex Ni[i-Pr(sal)Me][t-Bu(sal)H] was not observed in the system represented by Eq. 5.11. This implies there is no fission of the ligand at the weak link, the azomethine. The reactions are over in <3 min but more than 10^{-3} sec. Comment on the stereochemistry of the salicylaldimine complexes (VII) is necessary for the mechanism to be appreciated. Most such complexes are taking part in a rapid equilibrium between square planar and tetrahedral forms, which are very close in energy at ordinary temperatures. The equilibrium favors the paramagnetic tetrahedral form for various substituents, especially R groups with branching at the α carbon. Greatly reduced ligand exchange was observed by Chakravorty and Holm [13] when one components of the exchange was essentially square planar (diamagnetic) (Eq. 5.12). No exchange was observed in 60 min. Mossop and Lockhart,

$$Ni[t\text{-Bu(sal)}5Me]_2 + Ni[n\text{-Pr(sal)}5Me]_2 \rightarrow \text{slow exchange} \qquad 5.12$$

however, observed rapid exchange between a tetrahedral and a square-planar complex [14] but a slow exchange between tetrahedral complexes with bulky N substituents. It is possible that minor amounts of catalytic or deactivating impurities could affect the rate. There was no obvious stereoselectivity in this reaction when optically resolved R groups were used [15].

Aminotroponiminates scramble in the same way as the salicylaldimines [16] and it is to be expected that a whole series of complexes of the labile square-planar → tetrahedral type would scramble, and probably on central atoms other than nickel, for example, Co(II). Contact shifts provide a simple method of investigating these systems [13, 16], as do mass spectra [14].

III. β-Diketonates

A. *Lanthanides and Yttrium*

Lanthanide metal chelates ML_3 (L is β-diketone) were found to exchange ligands in the gas phase in a mass spectrometer, either with other lanthanide trischelates or with alkali metal β-diketonates. The reactions are given in 5.13 and 5.14, where M is a lanthanide metal and M′ is an alkali metal.

$$ML_3 + ML_3' \rightleftharpoons MLL_2' + ML'L_2 \qquad\qquad 5.13$$

$$ML_3 + M'L' \rightleftharpoons MM'L'L_3 + MM'L_2'L_2 \qquad\qquad 5.14$$

The systems are somewhat complicated in that the alkali metal β-diketonates are really trimeric in the mass spectrometer and can undergo self-exchange, viz. Eqs. 5.15 and 5.16. Alkali metal contamination can thus cause ligand

$$M_3L_3 + M_3L_3' \rightleftharpoons M_3LL_2' + M_3L'L_2 \qquad\qquad 5.15$$

$$M_3L_3 + M_3'L_3 \rightleftharpoons MM_2'L_3 + M_2M'L_3 \qquad\qquad 5.16$$

exchange in lanthanide chelates and its presence is to be avoided. The compounds $MM'LL_3'$ have been obtained as solids [17]. Other ligands such as 8-hydroxyquinoline were effective, thus sodium oxinate with holmium tris(trifluoroacetylacetonate) gave all possible exchange products in Eq. 5.14.

Since most of the lanthanide complexes are paramagnetic, NMR has not usually been helpful. However, yttrium is diamagnetic in oxidation state III. The eight-coordinated anions $Y(tfac)_4^-$ and $Y(hfac)_4^-$ (where tfac and hfac are β-diketone ligands) were studied by ^1H NMR in $CDCl_3$ solutions as their tetraphenylarsonium salts. The methine resonance of the β-diketone could be resolved [18] for each different environment of a specific ligand, viz.

$Y(tfac)_4^-$; $Y(tfac)_3(hfac)^-$; $Y(tfac)_2(hfac)_2^-$; $Y(tfac)(hfac)_3^-$; $Y(hfac)_4^-$.

Equilibrium constants were computed for accessible equilibria (5.17 and 5.18) at $-40°$.

$$Y(hfac)_4^- + Y(tfac)_2^-(hfac)_2 \rightleftharpoons 2Y(hfac)_3(tfac) \qquad\qquad 5.17$$

where $K = 4.9 \pm 0.64$,

$$Y(hfac)_3(tfac)^- + Y(hfac)(tfac)_3^- \rightleftharpoons 2Y(hfac)_2(tfac)_2^- \qquad 5.18$$

where $K = 4.0 \pm 0.5$.

Ligand exchange rates were equal for the two ligands but were too rapid for the mechanism to be discovered. Other eight-coordinate β-diketonate structures are known to engage in scrambling reactions (Section IIIB). See also Chapter 6, aluminum and gallium.

B. Zirconium, Hafnium, Cerium, and Thorium in Oxidation State IV

The eight-coordinate Zr chelates $Zr(acac)_4$ and $Zr(tfac)_4$ have been found to exchange ligands forming preferentially the mixed complexes. The reason for this preference is the increased entropy, because the reaction enthalpy is almost zero [19]. Pinnavaia and Fay [19] examined the system by ^{19}F NMR and obtained the constants in Table 5.3 for reactions 5.19–5.21.

$$Zr(acac)_4 + Zr(tfac)_2(acac)_2 \rightleftharpoons 2Zr(acac)_3(tfac) \qquad 5.19$$

$$Zr(tfac)(acac)_3 + Zr(tfac)_3(acac) \rightleftharpoons 2Zr(tfac)_2(acac)_2 \qquad 5.20$$

$$Zr(tfac)_2(acac)_2 + Zr(tfac)_4 \rightleftharpoons 2Zr(tfac)_3(acac) \qquad 5.21$$

These data are directly comparable with those for yttrium (Eqs. 5.17 and 5.18). The entropy is greater than that calculated statistically, possibly due to solvation effects. The same system has been studied using ^1H NMR by Adams and Larsen [20] and given the alternative description in Eqs. 5.22–5.24 and Table 5.4 which contains also data for the corresponding

$$3/4\ M(acac)_4 + 1/4\ M(tfac)_4 \rightleftharpoons M(acac)_3(tfac) \qquad 5.22$$

$$1/2\ M(acac)_4 + 1/2\ M(tfac)_4 \rightleftharpoons M(acac)_2(tfac)_2 \qquad 5.23$$

$$1/4\ M(acac)_4 + 3/4\ M(tfac)_4 \rightleftharpoons M(acac)(tfac)_3 \qquad 5.24$$

hafnium reactions. Although coalescence of spectra with increasing temperature was observed, the rates of reaction were not quantitatively studied. Nevertheless, a qualitative exchange order $Th > Ce > Zr$, Hf can be obtained from coalescence temperatures: 43° (Th), 60° (Ce), and >160° (Zr, Hf), although catalysis by decomposition products interferes too strongly for quantitative rate measurements.

TABLE 5.3

THERMODYNAMIC DATA FOR REACTIONS 5.19–5.21 IN BENZENE AT 0.25 M CONCENTRATION[a]

Eq.	Mean value of K^b	Statistical K	ΔH (kJ mole^{-1})	ΔS^c (J °K^{-1})	Statistical ΔS^c (J °K^{-1})
5.19	6.09 ± 0.21	2.67	0.2 ± 2.25	14.2 ± 7.54	8.18
5.20	6.08 ± 0.20	2.25	−0.2 ± 0.75	14.6 ± 2.9	6.74
5.21	7.23 ± 0.17	2.67	0.09 ± 1.12	16.3 ± 3.7	8.18

[a] From Pinnavaia and Fay [19].
[b] Temperature, 31°.
[c] Temperature, 25°.

TABLE 5.4

EQUILIBRIUM CONSTANTS OBTAINED FOR Zr AND Hf SYSTEMS DESCRIBED BY EQS. 5.22–5.24 IN BENZENE SOLUTION AT 33°[a]

Eq.	K_{Zr}	K_{Hf}	$K_{statistical}$
5.22	24	19	4
5.23	52	41	6
5.24	18	15	4

[a] From Adams and Larsen [20].

C. Uranium(IV)

Despite uranium(IV) being paramagnetic (two $5f$ electrons), proton NMR spectra can be obtained for its β-diketonate derivatives. The contact shifts cause the methide protons to appear at least 12 ppm downfield from tetramethylsilane. Siddall and Stewart [21] reported data in deuterio-chloroform at 30°. The simple tetrakis chelates (VIII), e.g., where R is

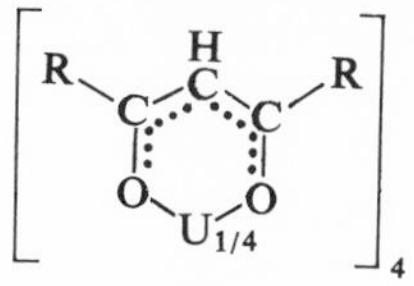

VIII

t-Bu, exhibit a single signal for the methide proton. Mixtures of two chelates (e.g., R = Et, R = t-Bu) show in addition separate signals for the three mixed chelates, but the curious feature is that these are very considerably shifted from those of the parent chelates (see Table 5.5), and the spread of shifts in the mixed chelates is, in the Et–t-Bu instance, over 9 ppm, while

TABLE 5.5

METHINE SIGNALS IN URANIUM TETRAKIS-β-DIKETONATES (VIII)[a]
FROM TETRAMETHYLSILANE (IN PPM)

Mixed chelates (R = t-Bu, Et)	VIII		Mixed chelates (R = t-Bu, Et)
	R = t-Bu	R = Et	
17.1	12.7	11.7	9.8
15.2			8.7
14.1			7.7

[a] From Siddall and Stewart [21].

for the parent chelates it is a mere 1 ppm. This suggests an exchange process, averaging a widespread set of shifts in the parent ions. Exchange between two or more distinct ligand sites would produce the observed spectra. Accordingly a rapid intramolecular process may be occurring (on the NMR time scale) while the intermolecular exchange of ligands is somewhat slower. When a diketone with perfluoroalkyl groups (R is C_nF_{n+2}) exchanges with a simple one, where R is Et, the range of shifts is in the region of 25 ppm.

IV. Miscellaneous Ligands

A. Substituted Pyridines on Nickel(II)

Forster et al. [22] have utilized contact shifts in a somewhat different way (see Section II) to determine the bonding preferences of nickel(II) for particular ligands. Competitive equilibria among pyridine and α-, β-, or γ-picoline, picoline oxide, triphenylphosphine, and triphenylphosphine oxide were studied for Ni(L)$_4$Cl$_2$ complexes, and some quantitative description of the system can be made [22]. The pyridine and β- and γ-picolines exchanged in almost random fashion, but the α-picoline was much less favored, possibly due to steric factors. The phosphine ligands are not favored in competition with pyridine.

B.   Alkoxyl, Dimethylamino, and Halide on Titanium(IV) and Zirconium(IV)

Dimethylamino and t-butoxide groups were scrambled on titanium in a second-order reaction in toluene; this was first order in both $Ti[NMe_2]_4$ and in $Ti[O-t-Bu]_4$. Rates were determined by following the appearance of the first scrambling product, $Ti(NMe_2)_3(O-t-Bu)$ for which the rate constant, $k_2 = 4.2 \pm 0.4 \times 10^{-5}$ liter mole^{-1} sec^{-1}.

The activation enthalpy was 40.4 kJ mole^{-1} and the enormous entropy of activation -197 J °K^{-1} was consistent with a four-center transition state, which should be considerably strained for the bulky ligands in this reaction. The rate is five orders of magnitude greater if the less bulky i-PrO group is used in place of t-BuO [23]. The alkoxides of titanium are usually polymeric in solution and most other studies of scrambling involving them show unusual heats of redistribution as a result.

A series of other substituents was scrambled on titanium (Eq. 5.25) but none was slow: the exchanges were in fact sufficiently rapid to cause

$$TiX_4 + TiY_4 \rightarrow TiX_3Y + TiX_2Y_2 + TiXY_3 \qquad 5.25$$

coalescence of proton NMR signals at 25°, thus preexchange lifetimes or limits to these could be obtained from NMR data for the exchange of MeO, EtO, i-PrO with Me_2N, MeO with t-BuO, MeO, EtO, i-PrO, and Me_2N with Cl (see Table 5.6) [24].

The equilibrium situation of the slow reaction of $(t$-BuO$)_4$Ti and $(Me_2N)_4$Ti was analyzed to provide the following K:

$$K_1 = [TiX_4][TiX_2Y_2]/[TiX_3Y]^2$$

$$K_2 = [TiX_3Y][TiXY_3]/[TiX_2Y_2]^2$$

$$K_3 = [TiX_2Y_2][TiY_4]/[TiXY_3]^2$$

where $K_1 = 0.24$, $K_2 = 0.40$, $K_3 = 0.11$, compared with the statistical values for $K_1 = K_3 = 0.375$ and $K_2 = 0.44$. The tertiary butoxide and dimethyl-aminotitanium species are monomeric and tetrahedral, probably because the bulk of the substituents prevents polymerization to give an increased coordination number on titanium. There is probably no association in the

TABLE 5.6

LIFETIMES OF EXCHANGE IN EQ. 5.25 AT 25°[a]

X	Y	τ (sec)
MeO	MeO[b]	0.13–0.16
MeO	Me$_2$N	~10^{-3}
Me$_2$N	Me$_2$N, OEt	0.05
t-BuO	MeO	2–8 × 10^{-2}
RO, Me$_2$N	Cl	10^{-3}

[a] From Weingarten and Van Wazer [24].
[b] The oligomer (MeO)$_4$Ti has four distinct chemical environments for MeO groups (distinguished by ^{1}H NMR).

mixed Me$_2$N–Cl or i-PrO–Cl species either. Here the enthalpies of mixing (of solutions in toluene) were measured and have very high values, the farthest from random so far observed [25, 26]. The thermodynamics of the other

TABLE 5.7

HEATS OF REACTION TiX$_4$ + TiY$_4$ (ΔH), EITHER FOR NEAT LIQUIDS OR FOR TOLUENE SOLUTION AT 25°[a]

X	Y	Mole ratio [Ti]/[X]	Concentration in toluene (g-atom Ti liter^{-1})	$-\Delta H$ [kJ (g-atom Ti)$^{-1}$]
EtO	Me$_2$N	0.5	3.3	16
i-PrO	Me$_2$N	0.5	Neat liquids	1.3
EtO	i-PrO	0.5	Neat liquids	19
EtO	t-BuO	0.5	Neat liquids	6.7
EtO	Cl	0.33	1.7	50
EtO	Cl	0.5	1.7	59
EtO	Cl	1.0	1.7	24
i-PrO	Cl	0.5	1.8	50
i-PrO	Cl	1.0	1.7	36
Me$_2$N	Cl	0.33	0.31	71
Me$_2$N	Cl	0.5	0.34	79.5
Me$_2$N	Cl	0.5	0.15	92[b]

[a] From Weingarten and Van Wazer [24].
[b] Precipitate observed but no correction for heat of precipitation made.

systems studied were complicated by the association of species; heats of redistribution were measured (as shown in Table 5.7) either for the mixing of neat liquids or of solutions in toluene. The enthalpies include heats of association and dissociation of the various bridged species in the reaction and an assumption of a realistic redistribution enthalpy [25] enables the heat of association of various bridged titanium compounds to be assessed. The structure of ethyl titanate has been determined by X-ray methods. It is a cyclic tetramer with two kinds of bridging alkoxy groups. The ^{1}H NMR spectra of the alkyl esters in this investigation can be interpreted to give rather different structures in solution, and it seems probable that the scrambling equilibria and kinetics would show temperature dependence reflecting the nature of the associated species.

The π-cyclopentadienyl compounds $(\pi\text{-cp})_2\text{TiClBr}$ and $(\pi\text{-cp})_2 \cdot \text{ZrClBr}$ were obtained in equilibrium with the appropriate dichlorides and dibromides in THF solutions after refluxing for about 5 hr. The equilibrium constants

$$K = [(\pi\text{-cp})_2\text{MXY}]^2/[(\pi\text{-cp})_2\text{MX}_2][(\pi\text{-cp})_2\text{MY}_2]$$

at 38° were 4.1 ± 0.6 [for $(\pi\text{-cp})_2\text{TiClBr}$] and 4.3 ± 0.2 for $(\pi\text{-cp})_2\text{ZrClBr}$. This should be compared with the expected random figure of 4 [27].

REFERENCES

1. R. J. Clark and E. O. Brimm, *Inorg. Chem.* **4**, 651 (1965).
2. J. P. Day, F. Basolo, and R. G. Pearson, *J. Amer. Chem. Soc.* **90**, 6927 (1968).
3. F. Basolo, *Chem. Brit.* **5**, 505 (1969).
4. R. J. Clark, *Inorg. Chem.* **3**, 1395 (1964).
5. C. A. Udovich, R. J. Clark, and H. Haas, *Inorg. Chem.* **8**, 1066 (1969).
6. W. J. Miles, Jr., and R. J. Clark, *Inorg. Chem.* **7**, 1801 (1968).
6a. R. C. Dobbie, private communication.
7. R. J. Clark, J. P. Hargaden, H. Haas, and R. K. Sheline, *Inorg. Chem.* **7**, 673 (1968).
8. C. A. Udovich and R. J. Clark, *Inorg. Chem.* **8**, 938 (1969).
9. F. Basolo, A. T. Brault, and A. J. Poë, *J. Chem. Soc.* **1964**, 676.
10. R. J. Clark and P. I. Hoberman, *Inorg. Chem.* **4**, 1771 (1965).
11. R. G. Hayter, *J. Amer. Chem. Soc.* **88**, 4376 (1966).
12. A. Davison, J. A. McCleverty, E. T. Shawl, and E. J. Wharton, *J. Amer. Chem. Soc.* **89**, 830 (1967).
13. A. Chakravorty and R. H. Holm, *J. Amer. Chem. Soc.* **86**, 3999 (1964).
14. W. J. Mossop and J. C. Lockhart, unpublished work.
15. R. E. Ernst, M. J. O'Connor, and R. H. Holm, *J. Amer. Chem. Soc.* **89**, 6104 (1967).
16. D. R. Eaton, D. J. Caldwell, and W. D. Phillips, *J. Amer. Chem. Soc.* **85**, 397 (1963).
17. J. R. Majer and R. Perry, *Chem. Commun.* **1969**, 271, 454.

18. F. A. Cotton, P. Legzdins, and S. J. Lippard, *J. Chem. Phys.* **45**, 3461 (1966).
19. T. J. Pinnavaia and R. C. Fay, *Inorg. Chem.* **5**, 233 (1966).
20. A. C. Adams and E. M. Larsen, *Inorg. Chem.* **5**, 228 (1966).
21. T. H. Siddall and W. E. Stewart, *Chem. Commun.* **1969**, 922.
22. D. M. Forster, K. Moedritzer, and J. R. Van Wazer, *Inorg. Chem.* **7**, 1138 (1968).
23. H. Weingarten and J. R. Van Wazer, *J. Amer. Chem. Soc.* **88**, 2700 (1966).
24. H. Weingarten and J. R. Van Wazer, *J. Amer. Chem. Soc.* **87**, 724 (1965).
25. J. C. Lockhart, *Chem. Rev.* **65**, 131 (1965).
26. K. Moedritzer, *Advan. Organometal. Chem.* **6**, 171 (1968).
27. P. M. Druce, B. M. Kingston, M. F. Lappert, T. R. Spalding, and R. C. Srivastava, *J. Chem. Soc. A*, **1969**, 2106.

6

Group III

The acceptor characteristics of Group III compounds are probably the most important factors in producing the redistribution reactions so well documented in this group, which are often rapid and random. For boron, aluminum, gallium, indium, and thallium compounds as unassociated three-coordinate monomers, four-center mechanisms are often proposed modeled on the structure of diborane, but for dimers the four-center transition state is often ruled out by virtue of being the starting structure. The dimerization of many aluminum and boron compounds is also responsible for somewhat different rate and thermodynamic characteristics than customary in redistribution.

I. Boron

Simple boron halides undergo mutual scrambling reactions on mixing and the mixed halides BX_2Y and BY_2X have all been identified either by ^{11}B [1] or by ^{19}F NMR [2, 3] and/or by mass spectra [3, 4] and infrared or Raman spectra [5–9]. The equilibrium is always approximately random, but is too rapidly attained in either direction for the mixed halides to be obtained pure. They can only exist in the redistribution equilibrium. Several attempts have been made to describe rates and equilibria in these systems quantitatively, but so far this has not had great success since the

[19]F nucleus seems to be the only convenient label on simple halides which can be used for accurate concentration measurements. The related organo-boron mixed halides, phenylboron chloride bromide [10, 11], ethyl-, methyl-, and vinylboron chloride fluorides [3, 11] and methylboron bromide fluoride [3] have all been identified by physical techniques. The halogen, but not the alkyl–aryl group, is labile in these compounds and again it is not possible to separate the mixed halides. A stable mixed halide, dimethyl-aminoboron chloride bromide was obtained on heating the corresponding dichloride and dibromide at 100°. The ^{11}B NMR spectrum of the mixture contained a new signal for the mixed compound. This could not be distilled out of the mixture, but was isolated as its dimer $(Me_2NBClBr)_2$ in crystalline form [12]. The diethylaminoboron halides (which do not dimerize) ex-changed halogen over a period of hours at room temperature. The rates of chloride–bromide exchange for a series of substituents at room temperature then seem to be $Me_2NBX_2 < Et_2NBX_2 < BX_3$ or $EtBX_2 < PhBX_2$ [12] with the appropriate half-lives of weeks (Me_2N) and hours (Et_2N), and estimated minimum lifetimes of $\geqslant 10^{-2}$ sec (BCl_3) and $\leqslant 10^{-2}$ sec $(PhBCl_2)$.

Although the simple halides scramble rapidly at room temperature, it is found that complexed halides (e.g., $Me_2O \cdot BX_3$ and $Me_3N \cdot BX_3$) are more stable. These compounds have convenient labels in the hydrogen nuclei of the ligand which can be used to distinguish the complexed mixed halide [13] in NMR spectra. No scrambling of halogen occurs even on heating to 180° when trimethylamine complexes are mixed in absence of free Lewis acid. It is possible that activation energies for halide exchange on complexed boron may be obtained. We can infer that free boron halide is necessary to cause scrambling, and that Lewis acidity is an essential feature of the mechanism. Although dimers of boron halides are not known under ordinary conditions, they are generally supposed to be intermediates in scrambling by four-center mechanisms; verification of their existence in a krypton matrix at 20°K was recently obtained [14].

The vibrational spectra of the mixed ternary halides (Cl, F; Cl, Br; and Br, F) were investigated by Lindemann and Wilson in gas and liquid phases [5]. The use of Raman and infrared spectra, together with $^{10}B–^{11}B$ isotope shifts, and band contours enabled assignment of all the observed frequencies for the mixed species which have C_{2v} symmetry. The data were in reasonable agreement with frequencies calculated using a simple VFF technique, where the parameters (force constant, bond length, and angle) employed for the mixed halides were assumed to be the same as for the homohalides. The fundamental frequencies observed in this study have been used in the estimation of thermodynamic properties for the mixed halides from equations derived from statistical mechanics (Chapter 2, Section I). Vibrational assignments have also been made for the mixed phenylboron bromide chloride [10] and calculations of thermodynamic quantities have been made [10].

There has been one attempt to measure the rate of a scrambling reaction of tris halides. The F–Cl reaction was found to have a half-life of ~10 min in the study by Nightingale and Crawford, who used a fast-scanning infrared spectrophotometer to follow the intensity changes in appropriate absorption bands [15]. An attempt to measure the rate of Cl–Br scrambling on the phenylboron residue was thwarted since the freezing points of the components prevented recording of ^{11}B NMR spectra at the necessary temperatures [10, 11].

Vibrational spectra have been used as measures of concentration for estimation of equilibrium constants in the F–Cl and Cl–Br scrambling reactions. Data are shown in Table 6.1. The other data in this table were

TABLE 6.1

EQUILIBRIUM CONSTANTS FOR SCRAMBLING HALOGEN ON BORON, OBTAINED
BY DIFFERENT PHYSICAL METHODS[a]

X	Y	K	Temp (°C)	ΔH (kJ mole^{-1})	Ref.
Cl	F	1.89[b]	27–29	7.01	8, 9
Cl	F	2.08[b]	15	4.6[e]	8
Cl	F	1.72[b]	40	5 ± 0.3[e]	8
Cl	F	0.63[c]	27		16
Cl	Br	0.15[d]			6
Cl	Br	0.11[d]			7

[a] $K = [BXY_2][BX_2Y]/[BX_3][BY_3]$.
[b] Infrared.
[c] Mass spectrometry.
[d] Raman.
[e] From ΔG value defined by $(-RT/n)(\ln K_{obs}) = \Delta H - T \Delta S$.

obtained by mass spectrometry. The parent ions BCl_3^+, BCl_2F^+, $BClF_2^+$, and BF_3^+ were detected by means of their different m/e ratios. The ion current due to each of these species was obtained at specific temperatures, and the constants K_{IC} (ion-current analogs of equilibrium constants) derived for these temperatures.

$$K_{IC} = [I_{BF_2Cl^+}]/[I_{BF_3^+}]^{2/3} [I_{BCl_3^+}]^{1/3}.$$

for formation of BF_2Cl, and

$$K_{IC} = [I_{BFCl_2^+}]/[I_{BF_3^+}]^{1/3}[I_{BCl_3^+}]^{2/3}$$

for formation of BCl_2F, where I is the ion current, K_{IC} = constant $\times K$, where K is the equilibrium constant for the reaction derived from partial pressures. A plot of $\log K_{IC}$ against $1/T$ has a slope independent of the constant, and equal to the enthalpy ΔH of the reaction [16]. (See Chapter 2, Section III.)

The mixed boric esters $B(OR)_{3-i}(OR')_i$ (where $i = 1$ or 2) have been prepared simply by mixing $B(OR)_3$ and $B(OR')_3$ in liquid, solution, and gas phases. To date the existence of mixed esters where R, R' are Me, Et; Me, n-Pr; Me, i-Pr; Me, n-Bu; Me, $ClCH_2CH_2$; Et, n-Pr; Et, n-Bu; n-Pr, n-Bu, is substantiated [17, 18]. None of them has been isolated because of the rapid disproportionation of mixed esters, but their presence has been detected either by ^1H NMR in solution or neat liquid phases [17] or by mass spectrometry in the gas phase [18]. Methyl hydrogen on mixed esters (except Me, Et) could be resolved adequately at 60 MHz for NMR analysis of concentrations, and the equilibrium constants for Eq. 6.1

$$B(OR)_3 + B(OR')_3 \xrightleftharpoons{K} B(OR)_2(OR') + B(OR)(OR')_2 \qquad 6.1$$

at 30° were R = Me, R' = n-Pr, $K = 1.25$; R = Me, R' = n-Bu, $K = 3.5$; R = Me, R' = $ClCH_2CH_2$, $K = 1.2$. The figures are not very accurate but indicate a distribution of alkoxy groups close to random (for which K would be 9), the stability of the tris esters being slightly greater than expected. Resolution of nuclear magnetic resonance spectra was not sufficiently good for quantitative thermodynamic data to be obtained for the other mixed esters mentioned. However, comparison of the mass spectra of the tris esters alone and admixed [18, 19] indicated the presence of the mixed esters in the mixtures (see Section III, Chapter 2). Exchange lifetimes were not calculated for any of these mixed ester systems, since there was more than one kind of line-broadening observed as temperature was varied in the range −40° to 80°.

Exchange of alkoxy groups on phenylboronate esters is probably faster than on boric esters, since duplicate methyl signals could not be obtained in mixtures of $PhB(OMe)_2$ and $PhB(O\text{-}i\text{-}Pr)_2$, but the single signal observed was chemically shifted from that of the pure dimethyl ester, indicating a preexchange lifetime for methyl groups <0.05 sec [17].

The reaction of the methyl ester of diisobutylborinic acid I with II the diethyl ester of isobutylboronic acid produces equilibria 6.2 and 6.3.

$$2i\text{-Bu}_2\text{B(OMe)} + i\text{-BuB(OEt)}_2 \;\rightleftharpoons\; 2i\text{-Bu}_2\text{B(OEt)} + i\text{-BuB(OMe)}_2 \qquad 6.2$$
$$\phantom{2i\text{-Bu}_2\text{B(OMe)}}\text{I}\text{II}$$

$$i\text{-BuB(OEt)}_2 + i\text{-BuB(OMe)}_2 \;\rightleftharpoons\; 2i\text{-BuB(OMe)(OEt)} \qquad 6.3$$

McCusker *et al.* [20] found random exchange of alkyl groups in mixtures of I and II in the mole ratio 2:1 using ^{1}H NMR as the detector. The lifetime of methoxy groups before exchange between $i\text{-Bu}_2\text{B}$— and $i\text{-BuB}$= sites was between 0.32 and 1.8 sec at room temperature.

Mixtures of methyl phenylboronate (3 moles) and methyl borate (2 moles) in toluene or nitrobenzene showed two sharp methyl signals which broadened as the temperature was raised but did not coalesce even at the solvent boiling point. Exchange lifetimes were calculated for this system, and an activation energy of 6–16 kJ mole^{-1}. This very low activation energy is perhaps credible in view of a similar figure from the work of Hofmeister *et al.* [21] on the B_2O_3–$(\text{MeO})_3\text{B}$ system. This scrambles to give a mixture of long-chain polyborates with end $(\text{MeO})_2\text{B}$—O—, middle —O(MeO)BO—, and branching

$$-\!\text{O}\!-\!\text{B}\!\!\begin{array}{c}\diagup\text{O}\\[-2pt]\diagdown\text{O}\end{array}$$

units and methyl borate. The half-life of methoxy groups before exchange in the system was 0.04 sec at room temperature and the activation energy was 18.5 kJ mole^{-1}.

The reaction of methyl phenylboronate with isopropyl borate [17] has $K = 1.0$, the statistical value (Eq. 6.4).

$$3\text{PhB(OMe)}_2 + 2\text{B(O-}i\text{-Pr)}_3 \;\overset{K}{\underset{\longleftarrow}{\longrightarrow}}\; 3\text{PhB(O-}i\text{-Pr)}_2 + 2\text{B(OMe)}_3 \qquad 6.4$$

The general inference from exchange lifetimes of alkoxy groups on boron is that rates of exchange are in the order $\text{PhB(OMe)}_2 >$ PhB(OMe)_2—$\text{B(OMe)}_3 > i\text{-BuB(OMe)}_2$. A similar order of substituent effects emerges from rate comparisons in the boron halide series also [12].

A bridged four-center transition state is often postulated for scrambling reactions in boron chemistry because of the existence of diborane and other bridged borane species. However diborane itself undergoes slow scrambling of bridged and terminal hydrogen, which has been monitored by using isotopes either of boron or hydrogen. Exchange of hydrogen and deuterium on diborane is of $\frac{3}{2}$ order and has an activation energy of 92 ± 13 kJ mole^{-1} [22]. The reaction order is consistent with a preequilibrium in which diborane dissociates to give a free monoborane (known only in the mass spectrometer) as a reactive intermediate. The rate step would then involve electrophilic attack of the BH$_3$ entity on a terminal diborane hydrogen with release of a new electrophile (BH$_3$ or BD$_3$). Exchange of ^{11}B and ^{10}B had a similar rate to D–H exchange at 25° [23] and it is reasonable to suppose the same mechanism would apply. A considerably faster process is known both for B$_2$H$_6$ and Me$_2$NB$_2$H$_5$, in which bridging and terminal hydrogen exchange in ether solutions [24, 25]. It has been stated that this exchange must be intramolecular since the coupling of six hydrogen (five for Me$_2$N case) to boron is retained in the fast exchange limit. This exchange is probably accelerated by ethers, causing a nucleophilic displacement on B of a bridge hydrogen, followed by rotation of the rest of the molecule around the remaining bridge (III).

III

Exchange of alkyl groups on boron is probably slow [26]. Some early work did suggest rapid scrambling, but it is possible this was due to catalysis by residual reagents. Alkylalanes are known to be effective catalysts for exchange of alkyl groups on boron [27, 28]. Boron hydrides are also catalysts for this exchange [27].

Isomerization of alkyl groups (secondary to primary, for instance) attached to boron occurs at the temperatures required for scrambling and the isomerization mechanism in some of these cases may be elimination of olefin to give a B—H bond (a), followed by hydroboration of the released olefin to give the isomeric alkyl group with boron attached to a different carbon in Eq. 6.5 [26, 29], where the dashes serve as labels for the carbons.

6.5

Such a path would be a convenient route for intermolecular scrambling of alkyl groups on boron, if in the hydroboration step (b) the olefin and the B—H come from different alkylboranes. The kinetics of the isomerization are first order for several systems which have been studied, and the rates are independent of solvent. Some activation entropies and energies are given in Table 6.2. The activation energies are all very similar, but the entropies

TABLE 6.2

Activation Parameters for the Isomerization of Certain Alkyl-
Boranes in the Temperature Range 120–150°C[a]

Isomerization	$E^{\ddagger}$ (kJ mole^{-1})	$\Delta S^{\ddagger}$ (J °K^{-1})
t-Bu-i-Bu$_2$B → i-Bu$_3$B	163 ± 4.6	$+59.5 \pm 10.9$
i-Pr$_3$B → n-Pr-i-Pr$_2$B	121.5 ± 6.3	-35.6 ± 14.7
i-Pr$_2$-n-Pr → n-Pr$_2$-i-PrB	121.5 ± 6.3	-38.9 ± 14.7
i-Pr-n-Pr$_2$B → n-Pr$_3$B	121.5 ± 6.3	-45 ± 14.7
sec-Bu-n-Bu$_2$B → n-Bu$_3$B	133 ± 0.8	-14.7 ± 2.1

[a] From Rossi *et al.* [29].

are very different. For the isomerization of *t*-butylisobutylborane to tris-isobutylborane the entropy is large and positive and for all the others it is negative. This suggests rather different mechanisms for different alkyl groups. The mechanism in Eq. 6.5, in which the olefin becomes detached forming two species, is consistent with positive activation entropy (the *t*-butyl isomerization), while an alternative mechanism (Eq. 6.6) has been suggested for those with negative entropy.

$$6.6$$

Halo-substituted boron hydrides have been utilized in the synthesis of diborane since the successful exploitation of Eq. 6.7 for this purpose by Schlesinger and Burg [30]. The monohalodiboranes B_2H_5X are known where X is Cl, Br, or I and they all undergo disproportionation similar to Eq. 6.7 [30–33]. The dihaloboranes are known for F, Cl, and Br, and they

$$6B_2H_5Cl \rightarrow 5B_2H_6 + 2BCl_3 \qquad 6.7$$

are all monomers: as tetratomic species they have received a great deal of attention from structural chemists and the vibrational spectra have frequently been recorded. Vibrational assignments are available for HBF_2 [34–38], $HBCl_2$ [30–42; see 32, 33] $HBBr_2$ [39, 43–45] and various deutero and boron-10 isotopic species. The frequencies calculated by the Green's function method are in good agreement with the experimental data [41]. Thermodynamic functions have been calculated for some of these species [35, 46] and the mean amplitude of vibration plus the thermodynamic functions from 200 to 2000°K have been tabulated [47]. The least moments of inertia for $HBCl_2$ and $HBBr_2$ have been obtained from rotational fine structure of fundamental vibrations [40, 48]. The microwave spectra of all the isotopic difluoroboranes have been recorded and the molecular parameters (bond lengths and angle) computed from rotational constants [49]. Nuclear magnetic resonance spectra of the dihaloboranes have been recorded [50–52]. There are many patents covering the synthesis of the B_2H_5X and HBX_2 compounds (not necessarily by scrambling) and their subsequent disproportionation reactions to give pure diborane [53–56].

The reaction of boron trifluoride with diborane to give difluoroborane is very slow in the absence of catalysts and there are by-products (Eq. 6.8). The related Cl, Br, and I reactions are faster as will be noted later. Pyrolysis at 250° to give a BF_2H–BF_3 mixture is convenient, since the back reaction of Eq. 6.8 is also slow [38]. Reaction of BF_3 with boroxine is considerably faster, the H of the HBF_2 formed coming from the OH groups [35, 57]. The exchange of dimethoxyborane with BF_3 is also a useful synthesis (Eq. 6.9). The ^{10}B isotope was used in these reactions to establish the boron source for the BF_2H. The exchange in equilibrium 6.9 may go through the

$$\tfrac{2}{3}BF_3 + \tfrac{1}{6}B_2H_6 \ \rightleftharpoons \ BF_2H \qquad\qquad 6.8$$

$$(MeO)_2BH + BF_3 \ \rightleftharpoons \ BHF_2 + (MeO)_2BF \qquad\qquad 6.9$$

$HB(OMe)_2BF_3$ complex since excess BF_3 is required. When $^{10}BF_3$ is used, very little ^{10}B appears in the BHF_2 product. Direct H–F exchange would seem not to occur; instead we get OMe–F exchange. Nadler and Porter's studies with the $(HBO)_3$–BF_3 reaction show that the boron in the final HBF_2 comes exclusively from the boroxine [58], so the mechanism must be different from the dimethoxyborane one.

Reaction 6.10 reaches isotopic equilibrium in 2–3 hr at room temperatures. The exchange of hydrogen between $H^{10}BF_2$ and deuteriodiborane is faster (6.11) being more than 95 % complete in 1 hr. The ^{10}B isotope is not diluted

$$BF_3 + H^{10}BF_2 \ \rightleftharpoons \ {}^{10}BF_3 + H^{11}BF_2 \qquad\qquad 6.10$$

$$H^{10}BF_2 + {}^{11}B_2D_6 \ \rightleftharpoons \ D^{10}BF_2 + B_2D_5H \qquad\qquad 6.11$$

in this reaction in 48 hr; therefore, H–F exchange must be much slower [38]. It has been suggested that a free BD_3 unit is not involved but a series of mutual displacements such as that in Eq. 6.12, which does permit H–D exchange without requiring H–F exchange. The enthalpy of reaction 6.8

$$F_2BH \quad + \quad \underset{D}{\overset{D}{>}}B\underset{D}{\overset{D}{<}}\underset{D}{\overset{D}{>}}B\underset{D}{\overset{D}{<}} \quad \rightleftharpoons$$

$$\rightleftharpoons \quad F_2BD \quad + \quad \underset{H}{\overset{D}{>}}B\underset{D}{\overset{D}{<}}\underset{D}{\overset{D}{>}}B\underset{D}{\overset{D}{<}} \qquad 6.12$$

has been calculated as $12.6 \pm$ kJ mole^{-1} [35] at 296°, while $K_{eq} = 0.49$.

Lynds and Bass have measured the equilibrium constant for the corresponding chloride reaction (Eq. 6.13) in the opposite direction, $K_p = 532 \pm 1$

$$6HBCl_2 \rightleftharpoons 4BCl_3 + B_2H_6 \qquad 6.13$$

atm^{-1} [46]. The half-life of the $HBCl_2$ was 30 min at 25°, and the reaction reached equilibrium after 168 hr. Myers and Putnam reported that $HBCl_2$ and B_2H_5Cl could be purified by low-temperature GLC and examined the decomposition of the purified materials. The reactions 6.14 and 6.15 were

$$5HBCl_2 \rightleftharpoons B_2H_5Cl + 3BCl_3 \qquad 6.14$$

$$B_2H_5Cl \rightarrow B_2H_6 + BCl_3 + BHCl_2 \qquad 6.15$$

observed at 30 and 35°, respectively. The half-life of B_2H_5Cl at 35° was about 25 hr. When completely free from impurities, it did not decompose in 1 hr at 25° or in 8 hr at 0° [33].

Cueilleron and Bouix [31] have examined the system again at ambient temperature and at 100° starting with varying amounts of B_2H_6 and BCl_3 and they report that at 100° with less than 10% B_2H_6 present, only $HBCl_2$ is found. In diborane-rich mixtures, however, B_2H_5Cl is the only chloroborane formed. They tabulated the equilibrium composition for various starting compositions of diborane and boron trichloride at room temperature, at 100° and again at room temperature after quenching from 100°. These are given in Table 6.3. Their general conclusions are that $BHCl_2$ is more stable in chloride-rich mixtures and at higher temperature, while B_2H_5Cl is more favored in hydride-rich mixtures and at low temperatures. All the possible equilibria in 6.13–6.15 can occur.

TABLE 6.3

EQUILIBRIUM COMPOSITION OF MIXTURES INITIALLY CONTAINING B_2H_6 AND BCl_3 AT VARYING RATIOS AND AT AMBIENT TEMPERATURES AND $100°$[a]

Initial diborane (mole %)	Temp (°C)	Molecules present at equilibrium			
		B_2H_6	BCl_3	$BHCl_2$	B_2H_5Cl
0–4	Room temp	✓	✓	—	—
	100	—	✓	✓	—
	Quenched from 100	—	✓	✓	—
4–10	Room temp	✓	✓	—	—
	100	—	✓	✓	—
	Quenched from 100	✓	✓	✓	—
10–20	Room temp	✓	✓	Trace	—
	100	✓	✓	✓	—
	Quenched from 100	✓	✓	✓	Trace
20–90	Room temp	✓	✓	✓	✓
	100	✓	✓	✓	✓
	Quenched from 100	✓	✓	✓	✓
90–98	Room temp	✓	Trace	Trace	✓
	100	✓	Trace	✓	Trace
	Quenched from 100	✓	Trace	Trace	✓
98–100	Room temp	✓	—	—	✓
	100	✓	—	✓	✓
	Quenched from 100	✓	—	—	✓

[a] From Cueilleron and Bouix [31].

Clearly these systems are connected by some delicate equilibria, and since they appear very susceptible to decompositions catalyzed by impurities, it may be that further studies will be necessary to unravel the very complicated threads. The industrial interest in this system suggests that there is considerable unpublished material which might clarify the situation completely. It is to be expected that the equilibria in the B–Br–H system are equally complicated and even more susceptible to impurities.

The compounds BH_2Cl, BH_2Br, $BHCl_2$, $BHBr_2$ and $BHBrCl$ have been obtained as their PH_3 adducts, which are stable to hydride–halide exchange on boron [58]. In the PH_3BHX_2 complexes an exchange process, possibly intermolecular H exchange, causes collapse of J_H coupling between hydrogen on the phosphine and hydrogen on borane in the ^1H NMR spectrum as the temperature is increased. It is faster when X is Cl than when X is Br. The mixed halide $PH_3BHBrCl$ exists in the redistribution equilibrium 6.16 at $-80°$, since fully resolved spectra for all three species are seen in a mixture initially equimolar in dibromide and dichloride only. At $30°$ the collapse of all three signals to a wide line has occurred. The exchange process responsible is probably halogen exchange. The order of collapse of signals in the system is as follows: the H_P–H_B coupling collapses first on the chloride, then on the mixed halide; coalescence of mixed halide with chloride signals occurs before the collapse of coupling in the bromide signal (Eq. 6.16), and finally coalescence of all three signals [58].

$$PH_3BHBr_2 + PH_3BHCl_2 \rightleftharpoons 2PH_3 \cdot BHBrCl \qquad 6.16$$

The compounds $BHCl_2$ and BH_2Cl have been reported as their diglyme and [59] dimethyl ether ($BHCl_2$ only) complexes [60]. Faulks and his co-workers used scrambling reactions to prepare the triethylamine complexes $Et_3N \cdot BH_2X$ (X is Cl, Br, or I) as solids (Eq. 6.17). These and the related complexes (X is Ph) were apparently stable to redistribution [61].

$$2Et_3N \cdot BH_3 + Et_3N \cdot BCl_3 \rightleftharpoons 3Et_3N \cdot BH_2Cl \qquad 6.17$$

$$6(MeO)_2BH \rightleftharpoons B_2H_6 + 4B(OMe)_3 \qquad 6.18$$

$$\begin{matrix} CH_2O \\ | \quad\quad BH \\ CH_2O \end{matrix} \rightleftharpoons \tfrac{1}{6}B_2H_6 + \tfrac{1}{3} \begin{matrix} CH_2O \\ | \quad\quad BOCH_2CH_2OB \\ CH_2O \end{matrix} \begin{matrix} OCH_2 \\ | \\ OCH_2 \end{matrix} \qquad 6.19$$

$$CH_2 \begin{matrix} CH_2O \\ CH_2O \end{matrix} BH \rightleftharpoons$$

$$\tfrac{1}{6}B_2H_6 + \tfrac{1}{3}CH_2 \begin{matrix} CH_2O \\ CH_2O \end{matrix} BOCH_2CH_2CH_2OB \begin{matrix} OCH_2 \\ OCH_2 \end{matrix} CH_2 \qquad 6.20$$

The alkoxyborane $(MeO)_2BH$ and the related 1,3-dioxaborolane and 1,3-dioxaborinane undergo disproportionation reactions 6.18–6.20 to give diborane and a trisalkoxy compound. For these disproportionation reactions, some equilibrium constants and rate constants have now been measured: the data are collected in Table 6.4. The free energies of the systems at $25°$ are very interesting, since they suggest very little thermodynamic difference between the five- and six-membered rings. The kinetic data are also very similar. Reasonable reaction paths for these (second-order) disproportionations assume a four-center transition state; one would *a priori* expect the five-membered ring to form this more readily due to the

TABLE 6.4

<small>Comparison of Disproportionation of 1,3-Dioxaborolane and 1,3-Dioxaborinane with Dimethoxyborane (see Eqs. 6.18–6.20)[a]</small>

Reaction	K_p (gas phase, 25°C)	ΔG (kJ mole^{-1})	k (rate constant for forward reaction)	Temp (°C)	Ref.
Eq. 6.18	63 ± 5	-0.41	1.6×10^{-6} mole^{-1} min^{-1} (heterogeneous)	60	62, 65
Eq. 6.19	0.38 ± 0.07	$+0.6 \pm 0.05$	$0.9 \pm 0.2 \times 10^{-4}$ mm hr^{-1}	25	63, 64
Eq. 6.20	10.9 ± 0.2	-1.4 ± 0.1	$2.2 \pm 0.3 \times 10^{-4}$ mm hr^{-1}	25	64

[a] Homogeneous conditions except where noted.[b]
[b] Data on heterogeneous systems also available [62–65].

relief of ring strain arising from such a step. However, the borolane is slightly more resistant to disproportionation [64].

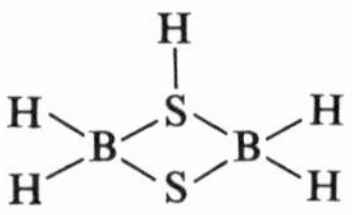

IV 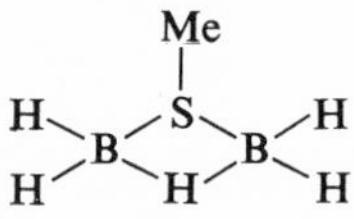   V

Several apparent scrambling reactions of μ-mercaptodiborane (IV) [66] and μ-methylthiodiborane (V) [67, 68] occur. Both compounds are thermally decomposed in other ways which compete with scrambling. Mercaptodiborane decomposes in the liquid phase approximately as follows in 6.21. The hydrogen in Eq. 6.21 has been shown to come in part from the S—H

$$2HSB_2H_5 \rightarrow H_2 + 1\% \, 2B_2H_6 + (1/x)(B_3S_2H_7)_x \qquad 6.21$$

group, since the specifically deuterated compound VI gave hydrogen of composition D_2 (4.3%), HD (68.4%), and H_2 (27.3%). No hydrogen or

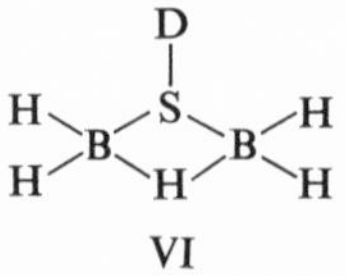

VI

methane was produced in the decomposition of μ-methylthiodiborane, which gave only thioborane polymers and diborane or trismethylthioborate [67, 68]. Probably the mechanism for H_2 production involves the S—H group. In the gas phase the decomposition of compound IV had a different stoichiometry from Eq. 6.21, less hydrogen being produced at low pressures. Exchange of the ^{10}B isotope between compound IV and $^{10}B_2H_6$ was found. It has been suggested that reaction takes place by prior dissociation of the diborane to give reactive $^{10}BH_3$, followed by electrophilic attack of BH_3 on the nucleophilic S bridge (as in VII).

VII	VIII	IX

The methylmercaptan–borane complex was found to decompose to give polymers of the approximate composition $(MeSBH_2)_x$, which, on reaction with diborane, gave compound V, which was very unstable, reverting to diborane and thioborane polymers at ambient temperatures (60% decomposition in 1 hr at 21°C) [68]. Other work has shown that polymeric alkylthioboranes decompose slowly at 25° to release trisalkylthioborates and $RSBH_2$ trimers (VIII) [67].

The ^{11}B NMR spectrum of μ-dimethylaminodiborane demonstrates the occurrence of an exchange process, since the expected coupling pattern (^{11}B to bridging and terminal H) for IX, a triplet of doublets, is observed at certain temperatures, while at higher temperatures, the pattern collapses to give a sextet, presumably because the five borane hydrogens of IX have become equivalent through exchange. This phenomenon has been studied kinetically in several solvents, and activation parameters are shown in Table 6.5. In methylcyclohexane there is probably a bridge-breaking mechanism with exchange of terminal and bridging hydrogen brought about by free rotation of the molecule around the remaining nitrogen bridge. In ether solutions, such a mechanism is accelerated by prior nucleophilic attack of ether on the boron to release the bridging hydrogen— compare the similar exchange in diborane, structure III [24]. The data in Table 6.5 indicate most striking differences between ether solvent and the nondonor solvent methylcyclohexane. High enthalpy and small positive entropy characterize the exchange in the nondonor solvent, while low enthalpy and large negative entropy terms characterize the two reactions in ether solvents. The entropy data in particular are appropriate for the mechanism suggested [25, 69].

Trimethylborane exchanges with diborane, and all the intermediate alkylboranes formed are dimeric with hydrogen bridges. These have been isolated generally by vacuum fractionation. Recent work has indicated that the infrared spectra of fresh fractions from vacuum distillation undergo

TABLE 6.5

INTRAMOLECULAR HYDROGEN EXCHANGE IN μ-DIMETHYLAMINODIBORANE: RATE CONSTANTS AND ACTIVATION PARAMETERS IN DIFFERENT SOLVENTS[a]

Solvent	Rate constant $(\sec^{-1})$ 0.5 M	$\Delta H^{\ddagger}$ (kJ mole^{-1})	$\Delta S^{\ddagger}$ (J °K^{-1})
Methylcyclohexane	14.5 (35°)	70.6 ± 0.42	1.7 ± 1.3
Tetrahydrofuran	9.4 (−50°)	25.6 ± 1.7	−107 ± 0.4
1,2-Dimethoxyethane		25.3 ± 4.2	−120 ± 1.7

[a] From Schirmer et al. [69].

rapid changes (in the space of about 80 sec at room temperature) which may indicate some rapid rescrambling process [70]. The early work of Schlesinger and Walker established the nature of the alkylborane products [71] and the equilibrium concentration of each; from this data, McCoy and Bauer have computed equilibrium constants for the systems in Eqs. 6.22–6.25 [72]. The disproportionation of tetramethyldiborane has been studied quantitatively at 0–39.4° [73].

$$2B_2H_5Me \rightleftharpoons B_2H_6 + B_2H_4Me_2, \quad K = 3.1 \qquad 6.22$$

$$3B_2H_4Me_2 \rightleftharpoons B_2H_6 + 2B_2H_3Me_3, \quad K = 5 \times 10^{-4} \qquad 6.23$$

$$4B_2H_3Me_3 \rightleftharpoons B_2H_6 + 3B_2H_2Me_4, \quad K = 3 \times 10^{-3} \qquad 6.24$$

$$6B_2H_5Me \rightleftharpoons 5B_2H_6 + 2BMe_3, \quad K \approx 10^{-12} \qquad 6.25$$

The redistribution equilibria of diborane as its solution in tetrahydrofuran with several substituted boranes (alkyl-, alkoxy-, halo-, and phenylthio-) have been studied using ^{11}B NMR measurements [74]. These reactions are all modified by complexing action of the solvent. Although borane is not present as a dimer (but as THF complex), some of the alkylboranes were dimerized. Some dimer-monomer equilibria were investigated for n-propyl- and isopropylboranes. The exchange between alkyl borates and diborane in tetrahydrofuran is singular in that the dialkoxy compound $(RO)_2BH$ is obtained but not the monoalkoxy compound $(RO)BH_2$. Disproportionation equilibria of some dialkoxyboron hydrides are presented in Table 6.6. The

TABLE 6.6

KINETIC AND THERMODYNAMIC DATA FOR THE REACTION OF DIALKOXYBORANES IN TETRAHYDRO-FURAN AT $25°C^a$

$$3(RO)_2BH \rightleftharpoons 2(RO)_3B + BH_3$$

R	Time required for equilibrium (days)	K
n-Pr	2	3.3
n-Bu	2	5.7, 6.1
i-Bu	7	3.2, 7.3
sec-Bu	18	3.8, 2.9
t-Bu	35	0.17, 0.13

^{a} From Pasto *et al.* [74].

relative rates of reaction for different alkyl groups shown in this table are consistent with a four-center bridged transition state, reached less easily by reagents with bulky R groups [74].

The reaction of trialkylboranes with boron trihalides to give mixed alkylboron halides requires high temperatures or catalysts. There is little quantitative information on these reactions although they have considerable synthetic use [75, 76].

Exchange of triphenylboron with trimethyl borate took place at 200°, but was nearly random. When the methyl borate was replaced by tris-dimethylaminoborane, exchange again occurred at 200°, but this time the diphenylboron compound was found to be particularly stable. Table 6.7 gives the appropriate equilibrium constants.

TABLE 6.7

EXCHANGE OF TRIPHENYLBORON WITH ALKOXY- AND DIALKYLAMINO-
BORANES[a]

	K (X = OMe)	K (X = NMe$_2$)
Ph$_3$B + PhBX$_2$ $\overset{K}{\rightleftharpoons}$ 2Ph$_2$BX	5.25	330
Ph$_2$BX + BX$_3$ $\overset{K}{\rightleftharpoons}$ 2PhBX$_2$	2	5

[a] From Hofmeister and Van Wazer [77].

TABLE 6.8

THERMOCHEMICAL DATA ON SCRAMBLING OF CHLORINE WITH OEt AND NMe$_2$ ON BORON

Reaction	ΔH (kJ mole^{-1}) (25°C)	Ref.
$\frac{1}{3}$BCl$_2$ + $\frac{2}{3}$B(OEt)$_3$ $\rightleftharpoons$ BCl(OEt)$_2$	16.7	78
$\frac{2}{3}$BCl$_3$ + $\frac{1}{3}$B(OEt)$_3$ $\rightleftharpoons$ BCl$_2$(OEt)	20.1	78
$\frac{1}{3}$BCl$_3$ + $\frac{2}{3}$B(NMe$_2$)$_3$ $\rightleftharpoons$ BCl(NMe$_2$)$_2$	37.2	79
$\frac{2}{3}$BCl$_3$ + $\frac{1}{3}$B(NMe$_2$)$_3$ $\rightleftharpoons$ BCl$_2$(NMe$_2$)$_2$	50.6	79

Exchange of halogen with alkoxy or dialkylamino groups on boron is
instantaneous, and the equilibrium favors the mixed compounds entirely
at room temperatures. Some thermochemical measurements are reported
in Table 6.8.

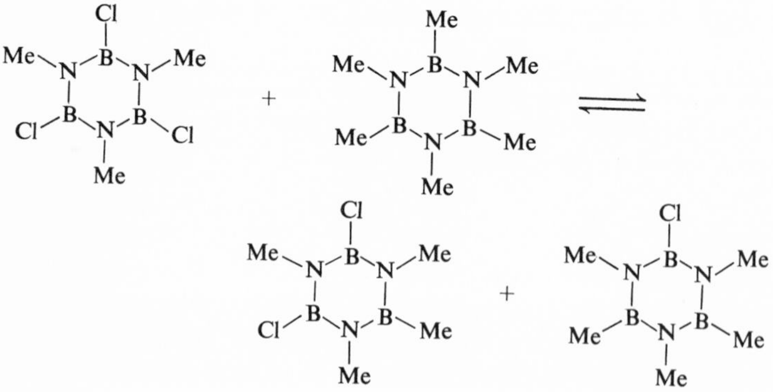

6.26

Scrambling of substituents on the boron atoms of borazoles occurs at elevated temperatures. At 175°, B—H bonds exchange with BMe, while between 250 and 350°, B—Cl and B—Me bonds interchange. The mixtures were analyzed by gas chromatography [80]. An equilibrium constant of 55 was obtained for Eq. 6.26, the expected random value being 9. Scrambling of B—H bonds of single borazoles with halogen on boron halides has also been used preparatively to produce new borazoles. A somewhat different kind of exchange reaction is the tetramer–trimer equilibrium of $(PhBN\text{-}i\text{-}Bu)_n$ borazynes. The tetramer was observed [81] to change smoothly to trimer at 250° (Eq. 6.27). This reaction was quantitative and irreversible.

$$3(PhBN\text{-}i\text{-}Bu)_4 \rightarrow 4(PhBN\text{-}i\text{-}Bu)_3 \qquad\qquad 6.27$$

II. Aluminum

A considerable number of exchange processes at aluminum have been investigated. The most extensive work is with organoaluminum compounds, where some fortunate circumstances have enabled features of several mechanisms to be established with detailed kinetic work. First, the "bridged dimer" often postulated as the transition state for a closed four-center mechanism is a stable species in aluminum chemistry; second, the proton resonance of the organic residue in alkylaluminum compounds exhibits chemical shifts in different environments of sufficient magnitude for kinetic work; and third, activation energies are sufficiently great that a dramatic coalescence of NMR signals with temperature can be observed. Some equilibrium constants are also available.

Much of this work stems from the observation [82] that a cyclopentane solution of methylaluminum (structure X) did not have separate proton resonances at room temperature for bridging (b) and terminal (t) methyl groups as would have been predicted for this known structure. Rapid exchange of bridging (b) and terminal (t) methyl groups has been advanced

$$\begin{array}{c}
Me_t \diagdown \diagup Me_b \diagdown \diagup Me_t \\
Al Al \\
Me_t \diagup \diagdown Me_b \diagup \diagdown Me_t
\end{array}$$

X

to explain this, because on cooling the solution, the alane proton NMR spectrum changes appropriately from one signal to two in the expected 1:2 ratio at −75°. An intramolecular [82, 83] mechanism (cf. diborane, III) and an intermolecular [84, 85] mechanism, Eqs. 6.28 and 6.29, have been propounded. In favor of the intramolecular mechanism, the energy for gas phase dissociation [86] of methylaluminum dimer (X) to monomer is 85.5 kJ mole^{-1}, while the activation energy for the methyl exchange process observed in the NMR spectrum was reported in the range 25–67 kJ mole^{-1} in different investigations [82–85]. The recent estimate [85] of 65 kJ mole^{-1} in cyclopentane is claimed to be quite accurate. Unfortunately, the energy of the dissociation to monomer is not known for the solution phase* in which the NMR measurements were performed, so that a firm negation of either route (1) or (2) to bridge–terminal methyl exchange is not possible. However, the analog trimethylgallane exists as a monomer, and its enthalpy of vaporization is known [87, 88]. If we can assume the value for monomer trimethylalane would be very similar in magnitude, we can write the cycle below and obtain a value of ΔH dissociation of X in liquid phase (d in the cycle); we then make the further approximation that the value for cyclopentane solution will not differ by more than a kilocalorie. This reduces the value of $\Delta H_{dissociation}$ from 85.5 to 65 kJ mole^{-1}, which is within the experimental error the same as the activation energy of the methyl exchange process [85]. This evidence is in favor of the intermolecular processes 6.28 and 6.29.

$$Al_2Me_6 \ (g) \quad \xrightarrow{\ a\ } \quad 2AlMe_3 \ (g)$$

$$b \uparrow \Delta H_{vap} \qquad\qquad c \downarrow -2\Delta H_{vap}$$

$$Al_2Me_6 \ (l) \quad \xrightarrow{\ d\ } \quad 2AlMe_3 \ (l)$$

$$a, \ \Delta H_{dissociation\,(g)} = 85.5 \pm 1.4 \ \ kJ\,mole^{-1} \ [86]$$

$$b, \ \Delta H_{vap} = 41.5 \pm 0.24 \ \ kJ\,mole^{-1} \ [87]$$

$$c, \ \Delta H_{vap} = 31 \ \ kJ\,mole^{-1}. \ [88]$$

There is also some strong kinetic evidence [85] for the second mechanism. The exchange of methyl groups between bridge and terminal positions in X has the same Arrhenius activation energy within experimental error as that for methyl exchange between X and the monomeric trimethylgallane or indane (Table 6.9). This is consistent with a rate-determining step for the latter exchanges not involving gallane or indane; since the rate depends on the ratio of monomer gallane or indane to dimer alane, but not on the

---

* The need for solution phase data for comparison here was pointed out to the author by M. E. Twentyman (personal communication, 1962).

$$Al_2Me_6 \underset{k_{-a}}{\overset{k_a}{\rightleftarrows}} 2AlMe_3 \underset{k_{-b}}{\overset{k_b}{\rightleftarrows}} AlMe_3 + AlMe_3 \qquad 6.28$$

X

<table>
<tr><td>solvent-caged
monomers</td><td>solvent-separated
monomers</td></tr>
</table>

$$AlMe_3 + GaMe_3^* \underset{k_{-c}}{\overset{k_c}{\rightleftarrows}} \begin{array}{c} Me \\ Me \end{array}\!\!Al\!\!\begin{array}{c} Me^* \\ Me \end{array}\!\!Ga\!\!\begin{array}{c} Me \\ Me \end{array} \rightleftarrows AlMe_3^* + GaMe_3$$

solvent-separated
monomers

solvent-caged
monomers

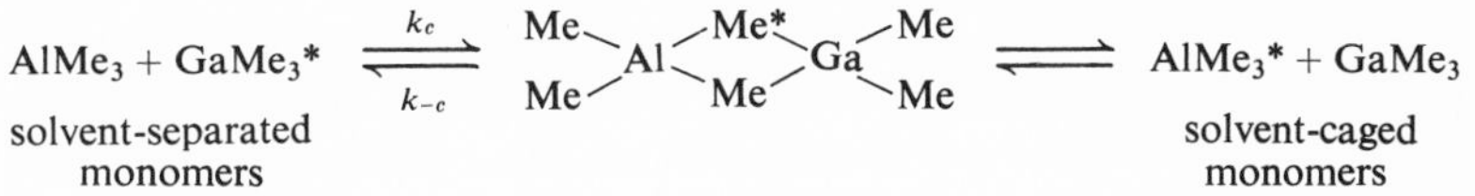

6.29

dilution, the mechanism suggested is that of reactions 6.28 and 6.29. The mean exchange lifetime of a methyl group on gallane is given by τ_E:

$$1/\tau_E = \tfrac{2}{3}\{k_a k_b k_c [Al_2Me_6]\}/((k_{-a} + k_b)\{2k_{-b}[AlMe_3] + k_c[GaMe_3]\})$$

In the conditions where $k_c[GaMe_3] \gg k_{-b}[AlMe_3]$ and $k_{-a} \gg k_b$, this reduces to $\tfrac{2}{3}\{k_a k_b [Al_2Me_6]/k_{-a}[GaMe_3]\}$, which fits the observed concentration dependence. In the limits where $k_c[GaMe_3] \gg k_{-b}[AlMe_3]$ and $k_b \gg k_{-a}$ (that is, where the solvent stabilizes the monomer and assists its separation from a solvent cage)

$$1/\tau_E = \tfrac{2}{3}\{k_a[Al_2Me_6]/[GaMe_3]\}$$

Other organoaluminum systems which have been studied are listed in Eqs. 6.30–6.37 and Tables 6.10 and 6.11, equilibrium constants K being quoted where known [89–96]. The asterisks in reactions 6.30–6.37 show

TABLE 6.9

Activation Energies for Exchange of Methyl Groups on Group III Metal Alkyls[a]

Exchange	$E^{\ddagger}$ (kJ mole^{-1})	Solvent
b-t-AlMe$_3$	64.5 ± 8	Toluene
GaMe$_3$–AlMe$_3$	66.5 ± 2	Cyclopentane
GaMe$_3$–AlMe$_3$	69 ± 4	Toluene
InMe$_3$–AlMe$_3$	68.7 ± 4	Toluene

[a] From Williams and Brown [85].

TABLE 6.10

EQUILIBRIUM CONSTANTS K FOR THE FORMATION OF MIXED DIMERS OF
ALUMINUM

$$\text{Me}_2\text{Al}\underset{Q}{\overset{Q}{<}}\text{AlMe}_2 + \text{Me}_2\text{Al}\underset{Z}{\overset{Z}{<}}\text{AlMe}_2 \underset{K}{\overset{\text{toluene}}{\rightleftharpoons}} 2\text{Me}_2\text{Al}\underset{Z}{\overset{Q}{<}}\text{AlMe}_2$$

Q	Z	K	Temp (°C)	Ref.
Me	PhC≡C	0.1	−(70–80)	94
Ph	PhC≡C	1	−(70–80)	95
Cl	Br	1	30	96
PhC≡C	i-PrO	2	80	96
Br	Ph	4	30	96
Me	Ph	5	−(70–80)	95
Me	p-CH$_3$C$_6$H$_4$	7	−(70–80)	95
Cl	PhC≡C	20	30	96
Br	PhC≡C	60	30	96
Br	i-PrO	100	80	97

$$2\text{PhMe}_2\text{Al}\cdot\text{OEt}_2 \underset{K=8}{\overset{\text{ether}}{\rightleftharpoons}} \text{Ph}_2\text{AlMe}\cdot\text{OEt}_2 + \text{Me}_3\text{Al}\cdot\text{OEt}_2 \qquad 6.30$$

$$\text{Me}_2\text{AlCl} + \text{Me*AlCl}_2 \underset{\text{or THF}}{\overset{\text{cyclopentane}}{\rightleftharpoons}} \text{Me}_2\text{*AlCl} + \text{MeAlCl}_2 \qquad 6.31$$

$$\text{Me}_6\text{Al}_2 + \text{Me}_2\text{Al}_2\text{Cl}_4 \overset{\text{ether}}{\rightleftharpoons} 2\text{Me}_4\text{Al}_2\text{Cl}_2 \qquad 6.32$$

$$\text{Me}_3\text{Al}\cdot\text{py} + \text{Ph}_3\text{Al}\cdot\text{py} \overset{\text{pyridine}}{\rightleftharpoons} \text{Me}_2\text{PhAl}\cdot\text{py} + \text{MePh}_2\text{Al}\cdot\text{py} \qquad 6.33$$

$$\text{Me}_6\text{Al}_2 + \text{Me}_4\text{*Al}_2\text{Br}_2 \overset{\text{anisole}}{\rightleftharpoons} \text{Me}_6\text{*Al}_2 + \text{Me}_4\text{Al}_2\text{Br}_2 \qquad 6.34$$

$$\text{Me}_6\text{Al}_2 + \text{Me}_4\text{*Al}_2\text{Cl}_2 \overset{\text{anisole}}{\rightleftharpoons} \text{Me}_6\text{*Al}_2 + \text{Me}_4\text{Al}_2\text{Cl}_2 \qquad 6.35$$

$$\text{Me}_6\text{*Al}_2 + \text{Me}_4\text{Ph}_2\text{Al}_2 \overset{\text{anisole}}{\rightleftharpoons} \text{Me}_6\text{*Al}_2 + \text{Me}_4\text{Ph}_2\text{Al}_2 \qquad 6.36$$

$$\text{Me}_4\text{Al}_2\text{Cl}_2 + \text{Me}_4\text{*Al}_2\text{Br}_2 \overset{\text{anisole}}{\rightleftharpoons} \text{Me}_4\text{*Al}_2\text{Cl}_2 + \text{Me}_4\text{Al}_2\text{Br}_2 \qquad 6.37$$

TABLE 6.11

ACTIVATION ENERGIES FOR SOME EXCHANGE REACTIONS OF ALUMINUM COMPOUNDS

Compounds		Concentration (M)	Donor concentration (M)	Solvent	$E^{\ddagger}$ (kJ mole^{-1})	Ref.
Me$_3$Al	EtMe$_2$Al	0.2	Ether, 0.2	Toluene	42	93
Me$_2$AlCl	Me$_2$AlBr	0.2	Ether, 0.2	Toluene	31	93
Me$_2$AlCl	Me$_2$AlBr	0.2	Anisole, 0.2	Toluene	56.5	93
Me$_3$Al	Me$_2$AlBr	0.2	Anisole, 0.2	Cyclopentane	65.0	93
Me$_3$Al	Me$_2$AlCl	0.2	Anisole, 0.2	Cyclopentane	66	93
Me$_3$Al	Me$_2$AlBr	0.2	Anisole, 0.2	Toluene	82	93
Me$_3$Al	Me$_2$AlPh	0.2	Anisole, 0.2	Toluene	53	93
Me$_3$Al	Me$_2$AlPhCC	—	—	Toluene	59	94

which methyls are involved in exchange. In these reactions the solvent has a very important role [91]. The Lewis acidity of the dimer alanes is such that they are present in donor solvents as monomer adducts with solvent. Exchange reactions are usually slowed down by donor solvent but the detailed mechanism for this depends largely on solvent and substituents.

Jeffery and Mole [92] have analyzed the mechanistic possibilities for exchange of groups Y between potentially dimeric aluminum compounds present as donor acceptor complexes $D \cdot AlY_3$ in donor solvents D and outlined four distinct situations:

(a) Bimolecular exchange of groups between two complexes with no loss of donor D before exchange.

(b) Rate-determining loss of D to give a highly reactive AlY_3 which undergoes subsequent and rapid exchange with $D \cdot AlY_3'$.

(c) Rapid preequilibria to give AlY_3 with subsequent rate-determining electrophilic attack on $D \cdot AlY_3'$.

(d) Rapid equilibria to give both AlY_3 and AlY_3' with a rate-determining reaction between them.

There are of course many other possibilities, but the processes (a)–(d), are intuitively simple, allowing direct fission of the weak donor–acceptor link, but a concerted process for fission of the stronger aluminum–alkyl link. Each process (a)–(d) should differ kinetically from the others. Kinetics for (a) would be first order in each reagent, second order overall; for (b) they would be first order in $DAlY_3$ only and inversely proportional to donor solvent concentration; for (c) they would be first order in each aluminum reagent and inversely proportional to donor solvent concentration; for (d) they would be first order in each aluminum reagent and inversely proportional to the second power of the donor solvent concentration. Thus far Jeffery, Mole, and co-workers have observed reactions whose kinetics are consistent with (a) or (c), or on the border line between the two. For strong donor–aluminum bonds one would expect mechanism (a), while in a series in which the donor–aluminum bond was progressively weakened, (b), (c), or (d) should become progressively more likely. This is borne out in studies of the exchange of methyl and phenyl groups on aluminum complexed with the series of donor solvents, pyridine, ether, anisole (decreasing order of donor strength). For pyridine [92] and ether [93] mechanism (a) has been observed, while for the weaker anisole, mechanism (c) is observed. Table 6.11 contains activation energies obtained for these reactions.

Reactions 6.34–6.37 have all been studied in anisole and mechanism (c) applies for 6.34–6.36. Reaction 6.37 differs in that anisole complexes are thermodynamically stronger: this leads to mechanism (a) being observed at high excess of anisole, and when diethyl ether is substituted for anisole, no inverse dependence on base concentration is observed [93]. The effect of pyridines on methyl–phenyl exchange on aluminum was to promote mechanism (a) except at very low excess of pyridine when (c) was thought to occur. The sterically hindered 2,6-lutidine probably leads to dissociative mechanism (c). Exchange of methyl between trimethylaluminum and dimethylphenylethynylaluminum in toluene proceeds with a half-life of about 10^{-1} sec at room temperature by the following mechanism [94]:

$$Me_6Al_2 \underset{}{\overset{rapid}{\rightleftharpoons}} 2Me_3Al$$

$$Me_3Al + Me_4(PhCC)_2Al_2 \underset{}{\overset{slow}{\rightleftharpoons}} \text{exchange products}$$

It is postulated that the transition state involves transfer of methyl between the two sites, with one aluminum five-coordinate:

$$\underset{Me}{\overset{Me}{\diagdown}}Al\underset{Me}{\overset{Me}{\diagup}}\underset{Me}{\overset{\overset{\displaystyle Me}{|}}{\diagdown}}Al\underset{CCPh}{\overset{PhCC}{\diagup}}Al\underset{Me}{\overset{Me}{\diagup}}$$

The activation energy was 59 kJ mole^{-1}, which may include a contribution of 0–30 kJ mole^{-1} for the production of Me$_3$Al [85]. (See Table 6.11.)

Dimer alanes with mixed bridges (e.g., tetramethyldialane chloride bromide, where the Cl and Br are bridging) have been made by scrambling reactions and their presence has been confirmed by NMR and mass spectra [97].

Mixed tetrahaloaluminates of the cation Me$_4$N$^+$ have been studied by means of ^{27}Al NMR in methylene halide solvents. Separate signals were observed for most of the possible mixed species (chloride, bromide, and iodide) when the mole ratio ([MX]/[AlX$_3$]) lay in the region 1.1–1.9. At lower ratios, exchange averaging of signals occurred. Both the ^{27}Al nuclear spin quantum number ($I = \frac{5}{2}$) and the lack of symmetry of the mixed species make the signals very broad (linewidths up to 57.6 ± 4.6 Hz). A monotonic increase of chemical shift is observed from AlCl$_4$$^-$ through the mixed halides to AlI$_4$$^-$ but this is not linear (see Chapter 10) [98].

The mixed complexes of aluminum (XI and XII) containing acetylacetone and dimethylformamide (DMF) have been characterized by their proton NMR spectra in DMF solvent [99].

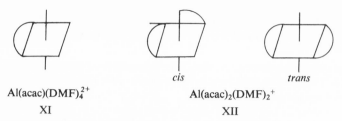

$\text{Al(acac)(DMF)}_4^{2+}$ $\text{Al(acac)}_2\text{(DMF)}_2^+$

XI   XII

Thermodynamic parameters for scrambling are given:

$$\text{Al(acac)}_3 + \text{Al(DMF)}_4\text{(acac)}^{2+} \rightleftharpoons 2\text{Al(acac)}_2\text{(DMF)}_2^+$$

where $K = 3.6 \pm 0.5$ (25°), $\Delta H = 20.6 \pm 2.1$ kJ mole^{-1}, and $\Delta S = 79.5 \pm 8$ J °K^{-1};

$$\text{Al(acac)}_2\text{(DMF)}_2^+ + \text{Al(DMF)}_6^{3+} \rightleftharpoons 2\text{Al(acac)(DMF)}_4^{2+}$$

where $K = 0.64 \pm 0.10$ (25°), $\Delta H = 32.2 \pm 4$ kJ mole^{-1}, and $\Delta S = 105$ J °K^{-1}. The equilibrium constant for the cis → trans isomerization of $\text{Al(acac)}_2\text{(DMF)}_2^+$ is 1.2 ± 0.1 and the first-order rate constant was 18 sec^{-1} at 5°. The rate of exchange of DMF between bulk solvent and the mixed complexes was estimated at 30 sec^{-1} at 5° for structure XII and 15° for structure XI.

TABLE 6.12

EQUILIBRIUM CONSTANTS K AT 25° FOR THE REACTION[a]

$$\text{M(AA)}_3 + \text{M(BB)}_3 \overset{K^b}{\rightleftharpoons} \text{M(AA)}_2\text{(BB)} + \text{M(AA)(BB)}_2$$

M	AA	BB	K
Al[c]	Acac	Tmhd	1.5
Al[c]	Acac	Hfac	3.34×10^4
Al[c]	Hfac	Tmhd	3.68×10^4
Ga[d]	Acac	Bzbz[e]	6.4
Ga[d]	Acac	Hfac	1.37×10^6

[a]From Fortman and Sievers [100] and Pinnavaia and Nweke [101].
[b] Statistical value = 9.
[c] Chlorobenzene.
[d] Benzene.
[e] Dibenzoylmethanate.

The ligands acetylacetonate (acac), tetramethylheptanedionate (tmhd), and hexafluoroacetylacetonate (hfac) exchange on aluminum. Equilibrium constants for some of the relevant exchange reactions are given in Table 6.12, with related data for some gallium compounds for comparison. Rates of exchange processes were also measured but there is some disagreement about the interpretation [100–102].

Aluminum alkyls, halides, and hydrides are all effective catalysts for the scrambling of substituents on other atoms, and are important in the formation of Ziegler catalysts for olefin polymerization. The Ziegler catalysts may be components of a scrambling reaction, but NMR studies have not been very successful in determining this. The paramagnetic transition metal ions broaden the NMR signals [90, 103] because of an unfavorable electron relaxation time.

III. Gallium, Indium, and Thallium

Information on scrambling of gallanes with lithium alkyls and aluminum alkyls will be found in Chapter 3, Section I and Section II of this chapter, respectively. Trimethylindane is discussed in Section II of this chapter.

The exchange of acetylacetonate for dimethylformamide ligands has been studied for gallium [104]. The exchanges 6.38 and 6.39 were found to take place in dimethylformamide solution at 25°, and the following thermodynamic data were obtained:

$$Ga(DMF)_6^{3+} + Ga(acac)_2(DMF)_2^+ \rightleftharpoons 2Ga(acac)(DMF)_4^{2+} \qquad 6.38$$

where $K_1 = 5 \pm 1$, $\Delta H_1 = 4 \pm 2$ kJ mole^{-1}, and $\Delta S_1 = 28 \pm 8$ J °K^{-1};

$$Ga(acac)_3 + Ga(acac)(DMF)_4^{2+} \rightleftharpoons 2Ga(acac)_2(DMF)_2^+ \qquad 6.39$$

where $K_2 = 74 \pm 7$, $\Delta H_2 = 1.3 \pm 4$ kJ mole^{-1}, and $\Delta S_2 = 38 \pm 4$ J °K^{-1}. Cis and trans isomers of Ga(acac)$_2$(DMF)$_2^+$ were present. The related exchanges for aluminum and beryllium have been discussed in the relevant sections of this book.

The equilibrium situation for exchange of diketonates on gallium resembles that for aluminum (Section II). Compare data for the two elements given in Table 6.12. Reactions with simple alkyl substituents on each diketone are approximately random, but if one ligand is fluorinated, the mixed compounds are strongly favored, this being mainly an enthalpy effect. Compare also the data on transition metal diketonates in Chapter 5, Section III.

Formation constants β_{ij} for the mixed species $InX_iY_j^{n+}$ and $TlX_iY_j^{n+}$, where $n = 3 - (i + j)$ or $1 - (i + j)$ and $X \neq Y =$ Cl, Br, and I, have been measured in aqueous solutions by polarographic and potentiometric techniques and solubility measurements [105]. Data are given in Table 6.13. Mixed tetrahalothallates ($X = Y =$ Cl, Br, or I) can be obtained as stable solid salts $Et_4NTlX_{4-n}Y_n$ ($n = 1, 2, 3$) and their vibrational spectra have been measured in acetonitrile showing the expected tetrahedral structure for the anions, with C_{2v} or C_{3v} symmetry [106]. Comparison of vibrational spectra for tetrahedral mixed halides is made in Chapter 2, Section I. There is no tendency for disproportionation in solid or solution [106].

TABLE 6.13

FORMATION CONSTANTS $\beta_{ij}{}^a$ FOR THE SPECIES $InX_iY_j^{n+}$ AND $TlX_iY_j^{n+}$, WHERE $n = 3 - (i + j)$ OR $1 - (i + j)$ AND $X \neq Y$ CAN BE Cl, Br, OR I^b

Species	β_{ij}	Species	β_{ij}
$TlBrCl^-$	6.3	$InCl_2^+$	316.2
$TlBr_2Cl^{2-}$	8.45	$InBrCl^+$	346.7
$TlBrI_2^{2-}$	172	$InBr_2^+$	33.1
TlI_2Br^{2-}	260	$InCl_3$	3548
		$InCl_2Br$	724.4
		$InBrCl_3^-$	794.3

^{a} $\beta_{ij} = [MX_iY_j]/[M][X]^i[Y]^j$, $K = [InBrCl^+]^2/[InCl_2^+]$-$[InBr_2^+] = 11.49$.
^{b} From Fridman et al. [105].

Exchange of alkyl and aryl groups on thallium has been investigated [107, 108]. Coupling between the two naturally occurring thallium isotopes (^{203}Tl, 29.5% and ^{205}Tl, 70.5% abundant), both of nuclear spin $\frac{1}{2}$, and the hydrogen nuclei in thallium alkyls is of the order of 150–250 Hz. The spin multiplets of trimethylthallium are observed at $-100°$ and collapse as the temperature is raised, implying exchange of methyl groups between thallium nuclei. (See Fig. 2.2.) This exchange is concentration and solvent dependent, probably second order. The activation energies were 25.2 ± 4.2 kJ mole^{-1} in methylene chloride, and 26.6 ± 2.1 kJ mole^{-1} in α-deuterotoluene. Mixed methyl–ethylthallium and methyl–vinylthallium compounds could be observed by ^{1}H NMR at low temperatures. Exchange in these systems and in triphenylthallium was considerably slowed by donor solvents (ether, trimethylamine). Such solvents would be expected to hinder any mechanism dependent on the Lewis acidity of the reagents, by forming adducts, e.g., $Me_3N \cdot TlPh_3$.

REFERENCES

1. P. N. Gates, E. F. Mooney, and D. C. Smith, *J. Chem. Soc.* **1964**, 3511.
2. T. D. Coyle and F. G. A. Stone, *J. Chem. Phys.* **32**, 1892 (1960).
3. F. E. Brinckman and F. G. A. Stone, *J. Amer. Chem. Soc.* **82**, 6235 (1960).
4. M. F. Lappert, J. B. Pedley, P. N. K. Riley, and A. Tweedale, *Chem. Commun.* **1966**, 788.
5. L. P. Lindemann and M. K. Wilson, *J. Chem. Phys.* **24**, 242 (1956).
6. J. Goubeau, H. J. Becher, and F. Griffel, *Z. Anorg. Allg. Chem.* **282**, 86 (1955).
7. D. Dollimore and L. H. Long, *J. Chem. Soc.* **1954**, 4457.
8. S. R. Gunn and R. H. Sanborn, *J. Chem. Phys.* **33**, 955 (1960).
9. T. H. S. Higgins, E. C. Liesegang, C. J. G. Raw, and A. W. Rossouw, *J. Chem. Phys.* **23**, 1544 (1955).
10. J. C. Lockhart, *Spectrochim. Acta* **24A**, 1205 (1968).
11. A. Finch and J. C. Lockhart, *Chem. Ind. (London)* **1964**, 497.
12. N. N. Greenwood and J. Walker, *Inorg. Nucl. Chem. Lett.* **1**, 65 (1965).
13. J. S. Hartman and J. M. Miller, *Inorg. Nucl. Chem. Lett.* **5**, 831 (1969).
14. J. M. Basler, P. L. Timms, and J. L. Margrave, *J. Chem. Phys.* **45**, 2704 (1966).
15. R. E. Nightingale and B. Crawford, *J. Chem. Phys.* **22**, 1468 (1954).
16. R. F. Porter, D. R. Bidinosti, and K. F. Watterston, *J. Chem. Phys.* **36**, 2104 (1962).
17. R. Heyes and J. C. Lockhart, *J. Chem. Soc. A*, **1968**, 326.
18. P. J. Fallon and J. C. Lockhart, *Int. J. Mass Spectrom. Ion Phys.* **2**, 247 (1969).
19. P. J. Fallon, P. Kelly, and J. C. Lockhart, *Int. J. Mass Spectrom. Ion Phys.* **1**, 133 (1968).
20. P. A. McCusker, P. L. Pennartz, and R. C. Pilger, *J. Amer. Chem. Soc.* **84**, 4362 (1962).
21. H. K. Hofmeister, J. R. Van Wazer, and K. Moedritzer, *Chem. Abstr.* **63**, 4402f (1965).
22. P. C. Maybury and W. S. Koski, *J. Chem. Phys.* **21**, 742 (1953).
23. I. Shapiro and B. Keilin, *J. Amer. Chem. Soc.* **77**, 2663 (1955).
24. D. F. Gaines, *Inorg. Chem.* **2**, 523 (1963).
25. D. F. Gaines and R. Shaeffer, *J. Amer. Chem. Soc.* **86**, 1505 (1964).
26. P. A. McCusker, F. M. Rossi, J. H. Bright, and G. F. Hennion, *J. Org. Chem.* **28**, 2889 (1963).
27. R. Koster, *Angew. Chem.* **75**, 1079 (1963).
28. R. Koster and G. Schomburg, *Angew. Chem.* **72**, 567 (1960).
29. F. M. Rossi, P. A. McCusker, and G. F. Hennion, *J. Org. Chem.* **32**, 450 (1967).
30. H. I. Schlesinger and A. B. Burg, *J. Amer. Chem. Soc.* **53**, 4321 (1931).
31. J. Cueilleron and J. Bouix, *Bull. Soc. Chim. Fr.* **1967**, 2945.
32. J. Cueilleron and H. Mongeot, *Bull. Soc. Chim. Fr.* **1967**, 1065.

33. H. W. Myers and R. F. Putnam, *Inorg. Chem.* **2**, 655 (1963).
34. L. Lynds and C. D. Bass, *J. Chem. Phys.* **43**, 4357 (1965).
35. R. F. Porter and S. K. Wason, *J. Phys. Chem.* **69**, 2209 (1965).
36. M. Perec and L. N. Becka, *J. Chem. Phys.* **43**, 721 (1965).
37. M. Perec and L. N. Becka, *J. Chem. Phys.* **44**, 3149 (1966).
38. T. D. Coyle, J. J. Ritter, and J. Cooper, *Inorg. Chem.* **7**, 1014 (1968).
39. L. Lynds, *Spectrochim. Acta* **22**, 2123 (1966).
40. L. Lynds and C. D. Bass, *J. Chem. Phys.* **40**, 1590 (1964).
41. T. Wolfram and R. E. De Wames, *Bull. Chem. Soc. Japan* **39**, 207 (1966).
42. C. D. Bass, L. Lynds, T. Wolfram, and R. E. DeWames, *J. Chem. Phys.* **40**, 3611 (1964).
43. L. Lynds, T. Wolfram, and C. D. Bass, *J. Chem. Phys.* **43**, 3775 (1965).
44. S. B. Rietti and J. Lombardo, *J. Inorg. Nucl. Chem.* **27**, 247 (1965).
45. S. K. Wason and R. F. Porter, *J. Phys. Chem.* **69**, 2461 (1965).
46. L. Lynds and C. D. Bass, *Inorg. Chem.* **3**, 1147 (1964).
47. G. Nagajaran, *Acta Phys. Pol.* **29**, 841 (1966); *Chem. Abstr.* **66**, 79769b (1967).
48. L. Lynds and C. D. Bass, *J. Chem. Phys.* **41**, 3165 (1964).
49. T. Kasuya, W. J. Lafferty, and D. R. Lide, *J. Chem. Phys.* **48**, 1 (1968).
50. T. D. Coyle, J. J. Ritter, and T. C. Farrar, *Proc. Chem. Soc.* **1964**, 25.
51. T. C. Farrar and T. D. Coyle, *J. Chem. Phys.* **41**, 2612 (1964).
52. T. Onak, H. Landesman, and I. Shapiro, *J. Phys. Chem.* **62**, 1605 (1958).
53. J. Cueilleron and J. L. Reymonet, French Pat. 1480, 303 (May 12 1967).
54. J. Cueilleron and J. L. Reymonet, U.S. Pat. 3,264,072 (Aug. 2 1966).
55. R. K. Pearson, U.S. Pat. 3,323,867 (June 6 1967).
56. J. L. Shepherd and T. C. Cromwell, U.S. Pat. 3,334,966 (Aug. 8 1967).
57. M. Nadler and R. F. Porter, *Inorg. Chem.* **6**, 1192 (1967).
58. J. E. Drake and J. Simpson, *J. Chem. Soc. A* **1968**, 974.
59. H. C. Brown and P. A. Tierney, *J. Amer. Chem. Soc.* **80**, 1552 (1958).
60. T. Onak, H. Landesman, and I. Shapiro, *J. Phys. Chem.* **62**, 1604 (1958).
61. J. N. G. Faulks, N. N. Greenwood, and J. H. Morris, *J. Inorg. Nucl. Chem.* **29**, 329 (1967).
62. H. S. Uchida, H. B. Kreider, A. Murchison, and J. F. Masi, *J. Phys. Chem.* **63**, 1414 (1959).
63. S. H. Rose and S. G. Shore, *Inorg. Chem.* **1**, 744 (1962).
64. G. E. McAchran and S. G. Shore, *Inorg. Chem.* **5**, 2044 (1966).
65. A. B. Burg and H. I. Schlesinger, *J. Amer. Chem. Soc.* **55**, 4020 (1933).
66. P. C. Keller, *Inorg. Chem.* **8**, 2457 (1969).
67. E. L. Muetterties, N. E. Miller, K. J. Packer, and H. C. Miller, *Inorg. Chem.* **3**, 870 (1964).
68. A. B. Burg and R. I. Wagner, *J. Amer. Chem. Soc.* **76**, 3307 (1954).
69. R. E. Schirmer, J. H. Noggle, and D. F. Gaines, *J. Amer. Chem. Soc.* **91**, 6240 (1969).
70. M. J. D. Low, R. Epstein, and A. C. Bond, *Chem. Commun.* **1967**, 226.
71. H. I. Schlesinger and A. O. Walker, *J. Amer. Chem. Soc.* **57**, 621 (1935).
72. R. E. McCoy and S. H. Bauer, *J. Amer. Chem. Soc.* **78**, 2061 (1956).
73. L. H. Long and M. G. H. Wallbridge, *J. Chem. Soc.* **1965**, 3513.
74. D. J. Pasto, G. Balasubramaniyan, and P. W. Wojtowski, *Inorg. Chem.* **8**, 594 (1969).
75. P. A. McCusker, G. F. Hennion, and E. C. Ashby, *J. Amer. Chem. Soc.* **79**, 5192 (1957).

76. W. Gerrard, "The Organic Chemistry of Boron." Academic Press, London, 1961.
77. H. K. Hofmeister and J. R. Van Wazer, *J. Inorg. Nucl. Chem.* **26**, 1209 (1964).
78. H. A. Skinner and N. B. Smith, *J. Chem. Soc.* **1954**, 3930.
79. H. A. Skinner and N. B. Smith, *J. Chem. Soc.* **1954**, 2324.
80. H. C. Newsom, W. G. Woods, and A. L. McCloskey, *Inorg. Chem.* **2**, 36 (1963).
81. B. R. Currell, W. Gerrard, and M. Khodabocus, *Chem. Commun.* **1966**, 77.
82. N. Muller and D. E. Pritchard, *J. Amer. Chem. Soc.* **82**, 248 (1960).
83. K. C. Ramey, J. F. O'Brien, I. Hasegawa, and A. E. Borchert, *J. Phys. Chem.* **69**, 3418 (1965).
84. C. P. Poole, H. E. Swift, and J. F. Itzel, *J. Chem. Phys.* **42**, 2576 (1965).
85. K. C. Williams and T. L. Brown, *J. Amer. Chem. Soc.* **88**, 5460 (1966).
86. A. W. Laubengayer and W. F. Gilliam, *J. Amer. Chem. Soc.* **63**, 477 (1941).
87. C. H. Henrickson and D. P. Eyman, *Inorg. Chem.* **6**, 1461 (1967).
88. L. H. Long and J. F. Sackman, *Trans. Faraday Soc.* **54**, 1797 (1958).
89. T. Mole and J. R. Surtees, *Aust. J. Chem.* **17**, 310 (1964).
90. Y. Sakurada, M. L. Huggins, and W. R. Anderson, *J. Phys. Chem.* **68**, 1934 (1964).
91. T. S. Mole, *Aust. J. Chem.* **18**, 1183 (1965).
92. E. A. Jeffery and T. Mole, *Aust. J. Chem.* **21**, 1497 (1968).
93. N. S. Ham, E. A. Jeffery, T. Mole, and J. K. Saunders, *Aust. J. Chem.* **21**, 659 (1968).
94. N. S. Ham, E. A. Jeffery, and T. Mole, *Aust. J. Chem.* **21**, 2687 (1968).
95. E. A. Jeffery, T. Mole, and J. K. Saunders, *Aust. J. Chem.* **21**, 137 (1968).
96. E. A. Jeffery, T. Mole, and J. K. Saunders, *Aust. J. Chem.* **21**, 649 (1968).
97. M. Fishwick, C. A. Smith, and M. G. H. Wallbridge, *J. Organometal. Chem.* **21**, P9 (1970).
98. R. G. Kidd and D. R. Truax, *J. Amer. Chem. Soc.* **90**, 6867 (1968).
99. W. G. Movius and N. A. Matwiyoff, *J. Amer. Chem. Soc.* **90**, 5452 (1968).
100. J. J. Fortman and R. E. Sievers, *Inorg. Chem.* **6**, 2022 (1967).
101. T. J. Pinnavaia and S. O. Nweke, *Inorg. Chem.* **8**, 639 (1969).
102. T. J. Pinnavaia, J. M. Sebeson, and D. A. Case, *Inorg. Chem.* **8**, 644 (1969).
103. E. N. Di Carlo and H. E. Swift, *J. Phys. Chem.* **68**, 551 (1964).
104. W. G. Movius and N. A. Matwiyoff, *Inorg. Chem.* **8**, 925 (1969).
105. Ya. D. Fridman, R. I. Sorochan, and N. V. Dolgoshova, *Russ. J. Inorg. Chem.* **7**, 1100 (1962).
106. R. A. Walton, *Inorg. Chem.* **7**, 1927 (1968).
107. J. P. Maher and D. F. Evans, *J. Chem. Soc.* **1963**, 5534.
108. J. P. Maher and D. F. Evans, *Proc. Chem. Soc.* **1961**, 208.

7

Group IV

Four-center mechanisms, which are dependent on the Lewis acidity of the central atoms, are probably important in many scrambling reactions of this group. In simple tetrahedral compounds of Group IV elements, Lewis acidity is not pronounced. It increases with atomic number Si $\ll$ Ge $\ll$ Sn [1]. This probably accounts for a like order of reaction rates for series of scrambling reactions of these elements. The slow reactions of silicon, often requiring days at high temperatures in the presence of catalysts, are particularly amenable to most physical techniques for study. One can isolate stable products and examine them at leisure. The heaviest contribution to Group IV redistribution chemistry has come from the work of Moedritzer and Van Wazer, who have fashioned NMR methods of analysis which have enabled them to determine the general bonding preferences of a great variety of groups with respect to silicon and germanium—these include monofunctional and many polyfunctional substituents. From their work, which is tabulated in detail in this chapter, it is now possible to predict equilibrium compositions for a wide range of systems and hence to select conditions favorable for production of polymers from systems with mixed central atoms, e.g., Si and Ge. This presages many new polymers for industrial scrutiny.

I. Silicon

A. Scrambling on a Single Residue

As is appropriate to the technical importance and terrestrial abundance of silicon, there is more information available about redistribution reactions at a central silicon atom than those at any other element. Halogens scramble on silicon on heating [2, 3] to give an approximately random distribution [2–4]. The concentrations of mixed halides present could be determined directly by fractionating the reaction mixture [4] and the Raman frequencies of the mixed halides have been used for their identification [2, 5]. An illustration of Raman spectra of the mixed silicon chloride iodides is shown in Fig. 2.1 (Chapter 2). Ester group interchange on silicon is also random (Table 7.1). Reaction between methyl and ethyl silicates in sealed glass tubes at

TABLE 7.1

EQUILIBRIUM CONSTANTS FOR THE SCRAMBLING REACTION AS DEFINED BY EQS. 7.1–7.3

Substituents						
X	Y	Temp (°C)	K_1 $(0.375)^a$	K_2 $(0.444)^a$	K_3 $(0.375)^a$	Ref.
OMe	OEt	150	0.417 ± 0.026	0.422 ± 0.025	0.386 ± 0.025	6
Et	n-Pr	180	0.694	0.315	0.454	7
Me	Cl	350	—	0.025	0.013	10
MeO	Cl	120	0.094 ± 0.019	0.042 ± 0.006	0.009 ± 0.003	6
Me_2N	Cl	25	6×10^{-4}	5×10^{-4}	6×10^{-6}	6

^{a} Values for the ideal random case.

150° took about 4 days to reach equilibrium [6]. The mixtures were then analyzed by GLC and the retention times of the mixed esters were assigned from their order of appearance between those of the pure tetrakis esters. Alkyl groups are known to scramble on silicon, but this is a very slow reaction. Tetraethyl- and tetra-n-propylsilanes were found to scramble randomly (Table 7.1) in less than 5 hr at 180°C in the presence of a powerful catalyst, aluminum chloride [7, 8]. The kinetics of catalyzed disproportionation of ethyltrimethylsilane in the presence of aluminum bromide have been measured. The rate was $\frac{3}{2}$ order in silane and first order in catalyst concentration, and had a very large entropy of activation (-189 J $°K^{-1}$) and an activation enthalpy of 58.6 kJ mole^{-1}. An S_F2 (four-center) mechanism is suggested by the entropy term [8].

The scrambling of alkoxy or dialkylamino groups with halogen on silicon is very strongly exothermic (Table 7.2). Thermochemical measurements led

TABLE 7.2

ENTHALPIES OF REDISTRIBUTION FOR THE REACTIONS
$(i/n)MX_n + (j/n)MY_n \rightleftharpoons MX_iY_j \quad (i + j = n)$

MX_iY_j	ΔH (kJ mole^{-1})	Ref.	MX_iY_j	ΔH (kJ mole^{-1})	Ref.
Si(OMe)$_3$(OEt)	0.071	6	Si(OMe)$_3$Cl	−16.0	6
Si(OMe)$_2$(OEt)$_2$	0.042	6	Si(OMe)$_2$Cl$_2$	−19.8	6
Si(OMe)(OEt)$_3$	0.218	6	Si(OMe)Cl$_3$	−15.9	6
Si(OEt)$_3$Cl	−18.0^a	9	Si(NMe$_2$)$_3$Cl	−32.6	6
Si(NMe$_2$)Cl$_3$	−27.2	6	Si(NMe$_2$)$_2$Cl$_2$	−38.1	6

^{a} Gas phase thermochemical measurement. All other values in this table are approximations based on δ $\Delta G = \Delta G_r - \Delta G_{stat} = (RT/n)/(\ln K/K_{stat}) \approx \Delta H_r$ [6].

indirectly to the value for the heat of the gas-phase reaction of tetraethyl silicate and silicon tetrachloride to give triethoxychlorosilane [10]. Nuclear magnetic resonance was used to analyze the composition of mixtures of tetrakisdimethylaminosilane or tetraethyl silicate with silicon tetrachloride [6]. The constants K for equilibria 7.1–7.3 are shown in Table 7.1. Enthalpies

$$SiX_2Y_2 + SiY_4 \rightleftharpoons 2SiXY_3 \qquad 7.1$$

$$SiX_3Y + SiXY_3 \rightleftharpoons 2SiX_2Y_2 \qquad 7.2$$

$$SiX_4 + SiX_2Y_2 \rightleftharpoons 2SiX_3Y \qquad 7.3$$

are given in Table 7.2. The composition diagram for exchange of Me$_2$N and Cl on silicon over the entire range is shown in Fig. 1c (Chapter 1). It is typical of reactions favoring the mixed compounds almost exclusively.

The OMe—Cl exchange was investigated kinetically in various solvents [11]. The rate of appearance of Si(OMe)$_3$Cl was followed by its distinctive ^{1}H NMR signal and gave acceptable rate constants up to about 30–40% reaction. The rate was first order in acid concentration of added (or adventitious) HCl, but at 100° the rate was less than that at 72°. An expression (Eq. 7.4) was found to fit the concentration dependence. The mechanism

$$d[Si(OMe)_3Cl]/dt = k_1 k_2[HCl][Si(OMe)_4]^2/\{k_{-1} + k_2[Si(OMe)_4]\} \quad 7.4$$

proposed to give this concentration dependence is shown in Eqs. 7.5 and 7.6.

$$\text{Si(OMe)}_4 + \text{HCl} \underset{k_{-1}}{\overset{k_1}{\rightleftharpoons}} \text{Si(OMe)}_4 \cdot \text{HCl (complex)} \qquad 7.5$$

$$\text{Si(OMe)}_4 + \text{complex} \underset{k_{-2}}{\overset{k_2}{\rightleftharpoons}} \text{Si(OMe)}_3\text{Cl} + \text{Si(OMe)}_4 + \text{MeOH} \qquad 7.6$$

The subsequent steps of the reaction, e.g., $\text{Si(OMe)}_3\text{Cl} + \text{SiCl}_4$, were found to have a similar rate dependence [11].

The exchange of alkyl and halogen on silicon is not random. Kinetics of this process have been studied in the presence of catalysts [12]. Trimethyl-silyl bromide exchanges in the presence of aluminum bromide, possibly through the intermediate complex (Eq. 7.7) which is known to form between the two. The reaction is first order in silyl bromide and aluminum bromide, so the proposed subsequent scrambling step (Eq. 7.8) must be fast [12].

$$\text{Me}_3\text{SiBr} + \text{Al}_2\text{Br}_6 \rightleftharpoons \text{Me}_3\text{SiBr} \cdot \text{Al}_2\text{Br}_6 \quad \text{(complex)} \qquad 7.7$$

$$\text{complex} + \text{Me}_3\text{SiBr} \rightarrow \text{Me}_4\text{Si} + \text{Me}_2\text{SiBr}_2 \cdot \text{Al}_2\text{Br}_6 \qquad 7.8$$

In general, the trends seen for scrambling on tetrafunctional silicon are also seen for scrambling on the moieties $\text{MeSi}\equiv$, $\text{Me}_2\text{Si}=$, and $\text{Me}_3\text{Si}-$, under conditions where the alkyl groups exchange too slowly to take part in the reactions. Effectively one can then study exchange on tri,- di-, and monofunctional silicon. These alkylsilicons share an attribute not possessed by tetrafunctional silicon, namely that NMR signals for the methyl ^{1}H nuclei can be observed. Moedritzer and Van Wazer and co-workers have made an impressive use of this tool. Chemical shifts of these methyl groups on silicon are found to be extremely sensitive to the nature of the other substituents on silicon, and mixed products are readily identified and analyzed. Although the thermodynamics are similar for the scrambling of any given pair of substituents on any specific moiety ($\text{Si}\equiv$, $\text{MeSi}\equiv$, $\text{Me}_2\text{Si}=$, or $\text{Me}_3\text{Si}-$), it is found that exchange of the same substituents between any two moieties (e.g., $\text{MeSi}\equiv$ and $\text{Me}_2\text{Si}=$) differs considerably from that on each specific moiety alone. The bonding preferences of each moiety for specific substituents are brought out clearly by scrambling studies on these systems.

Table 7.3 contains the equilibrium constants K_1 and K_2 for Eqs. 7.9 and 7.10 which adequately describe the equilibrium compositions of methyl-silicon systems ($\text{MeSi}\equiv$), for scrambling of various pairs of monofunctional substituents. Table 7.4 contains scrambling data for $\text{Me}_2\text{Si}=$ compounds, stated as the equilibrium constant K for Eq. 7.11. Moedritzer and Van Wazer have analyzed the data of Table 7.4, which describes the simplest type of

TABLE 7.3

Equilibrium Constants for Exchange of Substituents on Methylsilicon Residues

X	Y	$K_1{}^a$	$K_2{}^a$	Temp (°C)	Conditions	Ref.
Br	Cl	0.43	0.40	120	<20 hr	13
Br	I	1.43	0.52	120		13
OEt	OMe	0.26	0.32	150	<7 days	14
Cl	OMe	0.042	0.018	150	<12 hr	14
Cl	NMe$_2$	5×10^{-4}	5×10^{-4}	25	<3 hr	14
OMe	NMe$_2$	0.30	0.40	120	<40 hr	14
SMe	Cl	0.1	0.16	120	14–1 days	15
SMe	Br	0.041	0.086	120	7–4 days	15
Br	NMe$_2$	1.4×10^{-3}	1.6×10^{-4}	25	<1 hr, 120°C	15
Cl	OPh	0.221	0.183	150	160–100 days	16
Br	OPh	0.38	0.277	150	<27–3 days	16
Cl	NCO	0.51	0.34	25	75–20 hr	16
Br	NCO	0.64	0.57	120	100–50 hr	16
Cl	NCS	0.40	0.46	25	<120 hr	16
SMe	NCO	0.77	1.19	120	<26 days	16
SMe	NCS	0.24	0.40	120	<22 hr	16
NCO	NCS	0.36	0.39	25	<7 days	16
Cl	H	0.01	0.48	100	<23 hr^b	17

^{a} K_1 refers to Eq. 7.9; K_2 refers to Eq. 7.10.
^{b} Bu$_4$N$^+$Cl$^-$ catalyst.

TABLE 7.4

Exchange of Substituents on Me$_2$Si=

X	Y	K	Conditions required for equilibrium to be reached (°C)	Ref.
Cl	H	0.098	<23 hr 100 (Bu$_4$N$^+$Cl$^-$)	17
Br	OMe	0.0064	<23 hr, 25	18
Cl	SMe	0.178	<100 hr, 120	18
Br	SMe	0.049	<46 hr, 120	18
OMe	SMe	2.92	<24 hr, 120	18
CN	Cl	0.216	<3 hr, 25	19
NCS	SMe	0.244	<90 hr, 25	19
CNO	NCS	0.222	<100 hr, 25	19
SMe	NCO	~4.6	<7 days, 120	20
CN	NMe$_2$	~10^{-4}	<5 days, 72	20
Br	NMe$_2$	4.2×10^{-4}	<1 hr, 120	20
SMe	OAc	8.6	<17 days, 120	20
Cl	OAc	0.34	<8 days, 120	20
Br	OPh	0.14	<8 days, 150	20
Cl	OPh	0.16	<27 days, 150	20

$$2\text{MeSiX}_2\text{Y} \overset{K_1}{\rightleftharpoons} \text{MeSiX}_3 + \text{MeSiXY}_2 \qquad\qquad 7.9$$

$$2\text{MeSiXY}_2 \overset{K_2}{\rightleftharpoons} \text{MeSiY}_3 + \text{MeSiX}_2\text{Y} \qquad\qquad 7.10$$

$$2\text{Me}_2\text{SiXY} \overset{K}{\rightleftharpoons} \text{Me}_2\text{SiX}_2 + \text{Me}_2\text{SiY}_2 \qquad\qquad 7.11$$

redistribution reaction—that in which only one equilibrium constant is needed to describe the equilibrium situation. As a gauge of deviation from randomness, they use the term $\delta \Delta G = \Delta G_r - \Delta G_{\text{stat}}$, which can be obtained from Moedritzer and Van Wazer [20]

$$\delta \Delta G = (-RT/n)(\ln K_r/K_{\text{stat}}) \qquad\qquad 7.12$$

where n is the functionality of the central metal atom, in this case 2, and the subscript r denotes measured values for the reaction in question, and stat denotes values calculated for the statistical distribution of substituents. The function $\delta \Delta G$ is largely due to the enthalpy of redistribution. For this $\text{Me}_2\text{Si}{=}$ system, values of $\delta \Delta G$ (kJ mole^{-1}) are given in Table 7.5. Moedritzer and Van Wazer [20] have sought for linear free energy relationships in

TABLE 7.5

VALUES OF $\delta \Delta G$ FOR SCRAMBLING OF X AND Y ON $\text{Me}_2\text{Si}{=}$[a]

X	Y	$\delta \Delta G$ (kJ mole^{-1})	X	Y	$\delta \Delta G$ (kJ mole^{-1})
CN	NMe$_2$	11.3	NCO	NCS	0.16
Br	NMe$_2$	10.5	NCS	SMe	0.04
Cl	NMe$_2$	9.2	OMe	OEt	−0.42
Cl	Me	7.5	Cl	NCO	−0.42
Br	OMe	5.9	Cl	Br	−0.42
Cl	OMe	5.4	Cl	NCS	−0.42
CN	SMe	4.2	Cl	OAc	−0.42
Br	SMe	2.5	Br	NCO	−0.84
NCO	OMe	2.5	CN	NCS	−0.84
Cl	H	1.7	Cl	I	−1.7
Br	OPh	1.3	OMe	SMe	−4.2
Cl	OPh	0.84	NCO	SMe	−4.6
Cl	SMe	0.42	SMe	OAc	−5.9
Br	CN	0.42	OMe	H	−7.1[b]
OMe	NMe$_2$	0.42	F	H	−8.0[b]
Cl	CN	0.16			

[a] From Moedritzer and Van Wazer [20].
[b] These values refer to exchange on MePhSi= residues.

this series of data, attempting to relate $\delta \Delta G$ values with the nature of the exchanging substituents [20]. $\delta \Delta G$ values for a substituent X are plotted against the second substituent Y [20]. The ordering of substituents Y is arbitrarily chosen to give reasonable linear relations for as many substituents X as possible. The best curves which they obtained from this treatment fell naturally into two sets, the good donor substituents (NMe_2, SMe, OMe, etc.) giving curves showing the opposite tendency of curvature from those for halogen. This is qualitatively a most interesting development. Accumulation of more data on other substituents is called for in this exploration of free energy relationships, which might rationalize much of this branch of chemistry.

Baney and Shindorf have made a more successful approach to a linear free energy relationship, using the reaction 7.13 in carbon tetrachloride solution [22] to provide the data. Here they are concerned with the substituent effect of nonscrambling substituents on the reaction. The exchanging groups are the same throughout and the reference compound is trimethylsilyl. The equilibrium constant K for reaction 7.13 was measured by NMR

$$RR'R''SiCl + Me_3SiNMe_2 \; \overset{K}{\rightleftharpoons} \; RR'R''SiNMe_2 + Me_3SiCl \qquad 7.13$$

methods for a series of substituents R, R', and R'' and a plot of $\log K$ versus the sum of the Taft substituent constants $\sum \sigma^*$ for the R groups was linear, obeying the relation $\log K = 0.89 \sum \sigma^*$. Numerical data are given in Table 7.6, which is arranged in order of decreasing K values.

Scrambling on $PhMeSi=$ and $H_2Si=$ residues (without disturbing the incumbent substituents) has been investigated and the results are shown in Table 7.7. A particular point of the investigation of H–F scrambling on $PhMeSi=$ was the comparison afforded between gas, solution, and neat liquid. It is extremely useful to have such a comparison, since most scrambling reactions are run in condensed phases and equilibrium data refer to those. A close correspondence is noted in Table 7.7 for the three media, and one can have confidence that there is no major effect on enthalpy or entropy from preferential solvation of mixed or unmixed compounds. Scrambling of silanic hydrogen occurs at $100°$ with catalysts, e.g., Bu_4NF for F–H exchange [23].

B. Intersystem Scrambling on Me_nSi Residues

The equilibria 7.14, 7.15, and 7.16 describe the exchange of substituents between the moieties $MeSi\equiv$ and $Me_3Si—$, $Me_2Si=$ and $Me_3Si—$, and

TABLE 7.6

EQUILIBRIUM CONSTANTS FOR EQ. 7.13 AND A LINEAR FREE ENERGY
RELATIONSHIP[a]

Chlorosilane	K	$\Sigma\sigma^*_{RR'R''}$	$\delta \, \varDelta G$ (kJ mole^{-1})
Cl_2CHMe_2SiCl	38.2	—	−9.2
Ph_2HSiCl	27.6	1.94	−8.8
$MePh_2SiCl$	16.09	1.69	−7.1
$MePhViSiCl$	9.4	1.20	−5.9
$ClCH_2Me_2SiCl$	7.9	1.05	−5.5
$PhMe_2SiCl$	3.5	0.60	−3.4
$(Me_3SiO)_3SiCl$	3.1	0.54	−3
HMe_2SiCl	2.6	0.49	−2.5
$CF_3CH_2CH_2Me_2SiCl$	1.9	0.32	−1.67
$Me_2ViSiCl$	1.7	0.40	−1.3
Me_3SiCl	1.0	0	0
$i\text{-}Pr_3SiCl$	0.6	−0.57	+1.7
$n\text{-}Pr_3SiCl$	0.5	−0.35	+1.7
Random value	1		0

[a] From Baney and Shindorf [22].

TABLE 7.7

THERMODYNAMIC DATA FOR SCRAMBLING OF X AND Y ON RR′Si=

$$RR'SiX_2 + RR'SiY_2 \xrightleftharpoons{\;K\;} 2RR'SiXY$$

R	R′	X	Y	K	Temp (°C)	K	Temp (°C)	$\varDelta H$ (kJ mole^{-1})	Ref.
H	H	I	Br	2.3 ± 0.05	39	2.44 ± 0.05	66	−2.94[a]	21
H	H	Cl	Br	3 ± 0.3	39	—	—	—	21
H	H	I	Cl	6.7 ± 0.8	39	10 ± 1	66	+2.52[a]	21
Ph	Me	H	F	0.023 ± 0.01	100	—	—	+7.96[b]	23
Ph	Me	H	OMe	0.037 ± 0.007	100	—	—	+7.1[b]	23
Ph	Me	H	Cl	8.5 ± 0.3	100	—	—	−1.17[b]	23
Ph	Me	H	F	150 ± 1	25	75 ± 0.4	65.3	+14.5[a]	24
Ph	Me	H	F				84.7	+16.3[b,c]	24
Ph	Me	H	F				84.7	+16.8[b,d]	24
Ph	Me	H	F				40	+13.4[e]	23

[a] From Van't Hoff equation. [d] Gas phase.
[b] $\delta \, \varDelta G$. [e] Neat liquids, calorimetric measurement.
[c] Benzene solution.

$$\text{MeSiX}_3 + 3\text{Me}_3\text{SiY} \rightleftharpoons \text{MeSiY}_3 + 3\text{Me}_3\text{SiX} \qquad 7.14$$

$$\text{Me}_2\text{SiX}_2 + 2\text{Me}_3\text{SiY} \rightleftharpoons \text{Me}_2\text{SiY}_2 + 2\text{Me}_3\text{SiX} \qquad 7.15$$

$$2\text{MeSiX}_3 + 3\text{Me}_2\text{SiY}_2 \rightleftharpoons 2\text{MeSiY}_3 + 3\text{Me}_2\text{SiX}_2 \qquad 7.16$$

$\text{MeSi}\equiv$ and $\text{Me}_2\text{Si}=$. The appropriate intersystem equilibrium constants are given in Tables 7.8, 7.9, and 7.10, together with approximate times taken by the neat liquid system to reach equilibrium. For random exchange of

TABLE 7.8

INTERSYSTEM K FOR SCRAMBLING OF SUBSTITUENTS X AND Y ON MeSi AND Me₃Si (EQ. 7.14)[a,b]

X	Y	Minimum time required for system to reach equilibrium	Temp (°C)	K_I
H	Cl	Catalyst Bu₄N⁺Cl, 28 hr	100	$1 \pm 0.9 \times 10^{-8}$
NMe₂	Br	13 days	Room temp	$6 \pm 1 \times 10^{-12}$
SMe	Br	210 hr	Room temp	$4 \pm 3 \times 10^{-6}$
SMe	Cl	396 hr	Room temp	$7 \pm 2 \times 10^{-7}$
NMe₂	Cl	14 days	Room temp	$1 \pm 0.2 \times 10^{-10}$
OMe	Cl	19 hr	120	$7 \pm 3 \times 10^{-10}$
Br	Cl	39 hr	120	10.3 ± 1.4

[a] From Moedritzer and Van Wazer [15, 17].
[b] $K = [\text{MeSiY}_3][\text{Me}_3\text{SiX}]^3/[\text{MeSiX}_3][\text{Me}_3\text{SiY}]^3$.

TABLE 7.9

INTERSYSTEM K FOR SCRAMBLING OF SUBSTITUENTS X AND Y ON Me₃Si AND Me₂Si (EQ. 7.15)[a,b]

X	Y	Minimum time required for system to reach equilibrium	Temp (°C)	K_I
Br	Cl	16 hr	120	2.5 ± 0.2
OMe	Cl	17 hr	120	$5.3 \pm 1.7 \times 10^{-4}$
NMe₂	Cl	10 days	Room temp	$8.0 \pm 3.3 \times 10^{-4}$
H	Cl	Catalyst R₄N⁺, 128 hr	100	$5.6 \pm 3.4 \times 10^{-3}$
SMe	Br	45 hr	120	$3.5 \pm 0.3 \times 10^{-3}$
NMe₂	Br	2 hr	Room temp	$4.2 \pm 0.2 \times 10^{-4}$
SMe	Cl	137 hr	120	$9.8 \pm 1.8 \times 10^{-3}$

[a] From Moedritzer and Van Wazer [17, 25].
[b] $K = [\text{Me}_2\text{SiY}_2][\text{Me}_3\text{SiX}]^2/[\text{Me}_2\text{SiX}_2][\text{Me}_3\text{SiY}]^2$.

TABLE 7.10

INTERSYSTEM K FOR SCRAMBLING OF SUBSTITUENTS X AND Y ON MeSi AND Me$_2$Si
(EQ. 7.16)^{a,b}

X	Y	Minimum time required for equilibration	Temp (°C)	K_I
Cl	Br	240 hr	120	$1.95 \pm 0.35 \times 10^{-1}$
Br	Cl	240 hr	120	5.52 ± 0.62
OMe	Cl	51 hr	120	$1.02 \pm 0.50 \times 10^{-8}$
Cl	SMe	220 hr	120	$2 \pm 0.34 \times 10^{5}$
Cl	NMe$_2$	5 days	25	$4.75 \pm 2.86 \times 10^{7}$
H	Cl	Catalyst Bu$_4$N$^+$Cl$^-$, 23 hr	100	$1.0 \pm 0.6 \times 10^{-7}$

^{a} From Moedritzer and Van Wazer [17, 26].
^{b} $K = [\text{MeSiY}_3]^2[\text{Me}_2\text{SiX}_2]^3/[\text{MeSiX}_3]^2[\text{Me}_2\text{SiY}_2]^3$.

substituents between moieties, the constants K_I should be unity. There are very great deviations from random behavior; even for systems where halogen is scrambled (which are mostly random in Tables 7.3–7.5), a definite preference is found. The lightest halogen is found at equilibrium bound preferentially to the silane with fewest methyl groups. However, when halogen is in competition with a donor substituent such as Me$_2$N or MeO, the halogen (Cl or Br) is always found on the silane with fewest methyl groups. If the substituents Me$_2$N were bound to Si in exactly the same way, with the same strength of bond, regardless of the number of methyl groups on silicon, we would expect scrambling close to random. However, Si does have d orbitals which could act as π acceptors from the filled non-σ-bonding (lone pair) orbitals on the donor substituents Me$_2$N, MeO, MeS, etc., while methyl groups are expected to form σ bonds alone to silicon. The deviations from random behavior in Tables 7.8–7.10 imply that the bond energy approach based on σ-bond contributions is not justified for intersystem scrambling. The very large deviations are assumed to come from π-bonding contributions to bond energies. In general, the data in Tables 7.8–7.10 indicate faster reaction the greater the number of methyl groups on the silanes.

C. Polymeric Systems

Scrambling of substituents which are more than monofunctional sometimes leads to polymeric structures. Some examples from silicon chemistry are discussed here, in which the difunctional substituents —O—, —S—, and

—NMe— act as bridging ligands. If the silicon is present as dimethylsilicon residue, then linear dimethylsiloxane, -silthiane, or -silazane chains can be built up. MeSi≡ or Si≡ can give branching networks and three-dimensional structures.

1. LINEAR CHAINS

Moedritzer and Van Wazer have examined scrambling in the system formed by cyclic tetradimethylsiloxane, $(Me_2SiO)_4(A)$, and the neso* molecule dimethylsilyl dichloride, $Me_2SiCl_2(B)$, in various proportions. The expected equilibrium 7.17 in which linear chains are built up from Me_2SiClO— end units and $Me_2Si\big\langle{}^{O-}_{O-}$ middle units was observed.

$$Me_2SiCl(OSiMe_2)OSiClMe_2 \qquad Me_2ClSi(OSiMe_2)_3OSiClMe_2$$
$$C \qquad\qquad\qquad\qquad D$$

$$2 \text{ end units} \ \rightleftharpoons \text{ neso + middle unit} \qquad\qquad 7.17$$

Amounts of A and B equimolar with respect to $Me_2Si\equiv$ were reacted and rates of appearance and disappearance of species were monitored during the reaction and are informative as to the route taken in the polymerization. A polymer $(Me_2SiO)_n$, which maximized early in the reaction, contained most of the starting compound A, and gradually disappeared, forming lower oligomers of which the three-chain and five-chain C and D are present in greatest amounts at equilibrium. The equilibrium constant K for Eq. 7.17 describes the extent of polymerization in the mixture provided there are no cyclic oligomers present, but it does not of course indicate the amounts of each size of chain [27]. Siloxanes have little tendency to form rings, but when the bridging group is —S— or —NMe—, ring–chain equilibria also require to be taken into account. Values of K for Eq. 7.17 when the neso molecule is Me_2SiCl_2, $Me_2Si(OMe)_2$, or $Me_2Si(NMe_2)_2$, and the middle groups are provided by cyclic $(Me_2SiY)_n$ (where Y is —O—, —S—, or —NMe—) are given in Table 7.11 [27–29].

Certain naturally occurring acid clays have been used as scrambling catalysts to produce random distributions of dimethylsiloxanes, starting from $(Me_2SiO)_4$ and $(Me_3Si)_2O$. After $\frac{1}{2}$ hr at 80°, the chains formed by direct incorporation of the tetrasiloxane into the dimer (with four and eight middle groups) were present in relatively large amount [30], but the distribution of chain lengths randomized as reaction proceeded. Clays such as

---

* The term neso is bestowed by Van Wazer and co-workers on the smallest molecule in any family of compounds [27].

TABLE 7.11

EQUILIBRIUM CONSTANTS IN FAMILIES OF DIMETHYLSILICON
COMPOUNDS[a]

X	Y	K	Temp (°C)	Ref.
Cl	O	0.11	200	27
Cl	S[b]	0.13	200	28
Cl	NMe[b]	2.7×10^{-3}	120	29
MeO	O	0.24	200	27
MeO	S[b]	0.40	200	28
MeS	S[b]	0.20	200	28
Me$_2$N	O	0.50	200	27

[a] $K = [\text{ends}]^2/[\text{neso}][\text{middle}]$
$= [Me_2SiX(Y_{1/2})]^2/[Me_2SiX_2][Me_2Si(Y_{1/2})_2]$.
[b] In these systems ring–chain equilibria are also important.

kaolinite, halloysite, or montmorillonite were found to be good catalysts
for scrambling of silanic hydrogen with bridging oxygen on silicon [31].
There is a tendency for like ligands to accumulate on the same silicon, which
provides the drive for polymerization, as in Eq. 7.19. The decreasing

$$(Me_2HSi)_2O \xrightarrow{\quad 60\text{–}100°\quad} Me_2SiH_2 + Me_2HSiO(SiMe_2O)_xSiHMe_2 \qquad 7.18$$

reactivity of siloxanes with increased oxygen substitution on silicon, suggests
that the ligand exchange is acid catalyzed, and that an electrophilic attack
near silicon is involved.

Polymeric alkyl silicates have been generated by partial hydrolysis of
tetraalkoxides or alcoholysis of silicon tetrachloride in the presence of water.
A substantial activation energy for structural reorganization was found and
the equilibrium composition was studied by NMR spectroscopy. Near-
random distribution of building units occurs and gel point considerations
show that ring formation is of importance in this system [32].

2. RING–RING AND RING–CHAIN EQUILIBRIA

Mixtures of the compounds (Me$_2$SiS)$_3$ (I) and (Me$_2$SiNMe)$_3$ (V) on
equilibration give mixed rings containing —S— and —NMe— bridges
(III and IV) and the four-membered silthiane (II) [33]. Equilibrium constants observed for the system are shown in Table 7.12. Ring–chain as well
as ring–ring equilibria are important for bridging —S— and —NMe— on

TABLE 7.12

EQUILIBRATION OF FOUR- AND SIX-MEMBERED RINGS IN THE SILTHIANE AND SILAZANE
SERIES[a]

Equilibrium constant	At 25°C	At 120°C
$K = [I]^2/[II]^3$	$3.6 \pm 0.7 \times 10^{-3b}$	$13 \pm 0.2 \times 10^b$
$K = [I][III]/[IV]^2$	0.34 ± 0.02	0.32 ± 0.02
$K = [IV][V]/[III]^2$	0.90 ± 0.04	0.98 ± 0.05

[a] From Moedritzer and Van Wazer [33].
[b] Liter mole^{-1}.

silicon. A linear chain may redistribute to give a shorter chain with the
same end groups, but minus some middle groups which are excised to form

I   II   III

IV   V

a ring consisting only of middle groups. This equilibrium will be concen-
tration-dependent and can be described by a constant of the form K_{rc}:

$$K_{rc} = \frac{[\text{middle groups in } r\text{-cized rings}]}{(1 + D)\{V_2 + (R/2)(V_0 - V_2)\}} \left[\frac{\text{ends} + 2(\text{chain middles})}{2(\text{chain middles})} \right]^r \quad 7.19$$

where the square brackets indicate concentrations in terms of mole fractions
of building units, D is the dilution in volumes of solvent per volume of neat
sample, V_0 and V_2 are the molar volumes of neso and middle groups in
liters per mole, and R is the overall molar ratio of the monofunctional
substituent on the end group to that of the central moiety (here $Me_2Si=$).
Equation 7.19 was derived using stochastic graph theory [34], and there are
further equations for systems complicated by the presence of branching
chains. The constants relating rings to linear chains for bridging —O—,
—S—, or —NMe— on $Me_2Si=$ are given in Table 7.13.

TABLE 7.13

X	r	K_{rc}	Ref.
S	2	30	29
S	3	307	29
NMe	3	300	28
O	3	0.048	30
O	4	1.462	30
O	5	1.352	30

II. Germanium

Scrambling of alkyl groups on germanium is random, whether brought about by aluminum halide catalysis in mixtures of two different alkyls, or by reaction of germanium tetrachloride with equimolar quantities of two different alkyl Grignard reagents. A recent study of these systems developed GLC methods of analysis for the germanes [9]. Scrambling of other groups on germanes has been studied by NMR methods and is reported in Table 7.14 in terms of the equilibrium constants K_i for disproportionation, the quantity $\delta \Delta G$ (defined by Eq. 7.12), and in some instances the directly measured enthalpy ΔH [35–37]. The times required for complete exchange varied from fractions of a second at 25° for the OMe–Cl or NMe$_2$–Cl exchange, to days at 300° for Me–Cl exchange, but reactions were always faster than for the comparable silicon exchanges (see Section I). The thermodynamic patterns of behavior in the redistribution chemistry of germanium echo those for silicon, but there are invariably kinetic differences, germanium systems reacting faster than silicon.

The two equilibrium constants required for scrambling of two substituents X and Y on the MeGe entity as described in Eqs. 7.20 and 7.21 are given in Table 7.15. Scrambling on Me$_2$Ge= is described by Eq. 7.22, and the

$$2\text{MeGeX}_2\text{Y} \rightleftharpoons \text{MeGeX}_3 + \text{MeGeXY}_2 \qquad 7.20$$

$$2\text{MeGeXY}_2 \rightleftharpoons \text{MeGeX}_2\text{Y} + \text{MeGeY}_3 \qquad 7.21$$

$$2\text{Me}_2\text{GeXY} \rightleftharpoons \text{Me}_2\text{GeX}_2 + \text{Me}_2\text{GeY}_2 \qquad 7.22$$

TABLE 7.14

DATA FOR DISPROPORTIONATION OF TETRAHEDRAL GERMANIUM
COMPOUNDS WITH MIXED SUBSTITUENTS X AND Y ACCORDING TO THE
EQUILIBRIUM

$$2GeX_iY_{4-i} \underset{}{\overset{K_i}{\rightleftharpoons}} GeX_{i+1}Y_{3-i} + GeX_{i-1}Y_{5-i}$$

GeX_iY_{4-i}	$K_i{}^a$	$\delta \Delta G$ (kJ mole^{-1})	Ref.
$GeClF_3$	0.18	−1.68	35
$GeCl_2F_2$	0.34	−1.26	35
$GeCl_3F$	0.51	0.0	35
$GeCl(OMe)_3$	0.0085	−4.62	35
$GeCl_2(OMe)_2$	0.025	−3.4	35
$GeCl_3(OMe)$	0.05	−2.52	35
$GeCl(OMe)_3$	0.0088	−4.8	36
$GeCl_2(OMe)_2$	0.028	−3.4	36
$GeCl_3(OMe)$	0.089	−1.9	36
$GeCl(NMe_2)_3$	0.0002	−26^b	35
$GeCl_2(NMe_2)_2$	0.0003	−35^b	35
$GeCl_3(NMe)_2$	0.0007	−25^b	35
$Ge(OMe)(NMe_2)_3$	0.40	0.0	35
$Ge(OMe)_2(NMe_2)_2$	0.08	−2.10	35
$Ge(OMe)_3(NMe)_2$	0.019	−3.78	35
$GeMe_3Cl$	0.02	−7.1	35
$GeMe_2Cl_2$	0.005	−10.9	35
$GeMeCl_3$	~0.001		35
$Ge(OMe)_3Br$	0.013	−4.3	36
$Ge(OMe)_2Br_2$	0.155	−1.34	36
$Ge(OMe)Br_3$	0.099	−1.68	36

appropriate equilibrium constants are given in Table 7.16. Again the
magnitudes of equilibrium constants are similar to those for the related
silicon systems, but the reaction times are shorter [38–43].

Intersystem exchange of groups on MeGe≡ and Me$_2$Ge= has been
studied and the intersystem equilibrium constants K_I are given in Table 7.17
[40]. These are similar in magnitude to the corresponding K_I for silicon
systems (Table 7.10). For scrambling of halogen versus a donor group
such as OPh, SMe there is a very strong preference of the Me$_2$Ge= group
for the halogen. There is a minor difference in behavior between the silicon
and the germanium system with respect to exchange of two halogens.
Me$_2$Ge has a bonding preference for the lightest halogen of a pair, whereas
with the corresponding silicon system, the Me$_2$Si residue bonds preferen-

TABLE 7.14 *(continued)*

GeX_iY_{4-i}	$K_i{}^a$	$\delta\ \Delta G$ (kJ mole^{-1})	Ref.
Ge(OMe)$_3$I	0.054	−0.7	36
Ge(OMe)$_2$I$_2$	0.055	−0.62	36
Ge(OMe)I$_3$	0.016	+0.04	36
Ge(OMe)$_3$NCO	0.012	+0.92	36
Ge(OMe)$_2$(NCO)$_2$	0.015	+0.5	36
Ge(OMe)(NCO)$_3$	0.035	(+1.34)	36
Ge(SMe)$_3$Cl	0.13	−1.72	37
Ge(SMe)$_2$Cl$_2$	0.11	−2.2	37
Ge(SMe)Cl$_3$	0.04	−3.5	37
Ge(SMe)$_3$Br	0.03	−3.94	37
Ge(SMe)$_2$Br$_2$	0.07	−2.94	37
Ge(SMe)Br$_3$	0.12	−1.85	37
Ge(SMe)$_3$I	0.78	−2.60	37
Ge(SMe)$_2$I$_2$	0.065	−3.5	37
Ge(SMe)I$_3$	0.08	−2.5	37
Ge(SMe)$_3$(OMe)	0.87^c	+1.38	37
Ge(SMe)$_2$(OMe)$_2$	0.4^c	−0.12	37
Ge(SMe)(OMe)$_3$	1.74^c	+2.53	37
Ge(OMe)$_3$SMe	1.24^d	+2.0	37
Ge(OMe)$_2$(SMe)$_2$	0.58^d	+0.4	37
Ge(OMe)(SMe)$_3$	0.87^d	+1.39	37

^{a} Random values $K_1 = K_3 = 0.375$; $K_2 = 0.444$.
^{b} Calorimetric values −34, −38, and −23.5 kJ mole^{-1}, respectively.
^{c} Calculated from ^{1}H NMR signals of MeS groups.
^{d} Calculated from ^{1}H NMR signals of MeO groups.

tially to bromine rather than chlorine [25, 40]. The enormous differences from random (for which K_I would be 1) have been attributed to π-bonding effects.

Scrambling of three or four different substituents (Cl, Br, I, OPh) on MeGe≡ moieties has been recorded for equilibria which are close to random so that appreciable concentrations of all species could be observed by NMR methods. Half-lives of the order of minutes at 30° were observed in these systems [39]

Tri-*N*-methylhexachlorocyclogermazane undergoes substituent exchange in mixtures with the neso molecule GeCl$_4$ [44]. Nuclear magnetic resonance has been used to measure equilibrium constants for the formation of polymeric germazanes. The system has also been generated by dehydrochlorination of mixtures of methylamine and germanium tetrachloride.

TABLE 7.15

EQUILIBRIUM CONSTANTS CALCULATED FROM EQS. 7.20 AND 7.21 FOR SCRAMBLING OF
X AND Y ON MeGe≡

X	Y	K (Eq. 7.20)	K (Eq. 7.21)	Temp (°C)	Ref.
Cl	Br	0.41	0.42	Room temp	38
Br	I	0.44	0.42	Room temp	38
Cl	I	0.72	0.79	Room temp	38
Cl	OPh	0.14	0.24	33	39, 40
Br	OPh	0.17	0.37	33	39, 40
I	OPh	0.99	1.32	33	39, 40
Cl	OMe	0.0132	0.021	25	41
Cl	SMe	0.048	0.04	120	41
Br	OMe	0.021	0.04	25	41
Br	SMe	0.031	0.032	120	41
I	OMe	0.106	0.15	25	41
I	SMe	0.043	0.034	120	41

Hydrogen chloride is precipitated as methylammonium chloride and equilibrium attained in the remnant mixture. For the initial mole ratio $[MeNH_2]/[GeCl_4] = 3:1$, the main product is cyclic trimer. At very low ratios of amine, $Cl_3GeNMeGeCl_3$ is the Ge—N compound formed. Large excesses of amine give soluble species with monofunctional MeNH groups

TABLE 7.16

EQUILIBRIUM CONSTANTS FOR EQ. 7.22 DESCRIBING THE Me$_2$Ge
SYSTEM

X	Y	K	Equilibrium conditions (°C)	Ref.
SMe	Cl	0.055	100 hr, room temp	42
SMe	Br	0.025	30 hr, room temp	42
SMe	I	0.0123	2 hr, room temp	42
Cl	Br	0.30	33	43, 40
Br	I	0.35	33	43, 40
Cl	OPh	0.14	33	40
Br	OPh	0.24	33	40
I	OPh	1.61	33	40
Br	OMe	0.011	33	38
Br	SMe	0.014	33	40
Cl	I	0.67	33	43, 40

TABLE 7.17

INTERSYSTEM K FOR SCRAMBLING OF SUBSTITUENTS X AND
Y ON MeGe$\equiv$ AND Me$_2$Ge$=$ AT 33°C[a, b]

X	Y	K
Cl	Br	5.92 ± 0.42
Cl	OPh	7.07×10^5
Br	I	3.4×10^2
Br	OPh	1.9×10^5
Br	OMe	2.1×10^9
Br	SMe	5.0×10^8
I	OPh	1.5×10^2
I	Cl	0.58×10^{-3}

[a] From Moedritzer and Van Wazer [40].
[b] $K = [\text{MeGeY}_3]^2[\text{Me}_2\text{GeX}_2]^3/[\text{MeGeX}_3]^2[\text{Me}_2\text{GeY}_2]^3$.

TABLE 7.18

EQUILIBRIUM CONSTANTS FOR CYCLIC AND LINEAR OLIGOMERS AND POLYMERS BASED ON
GERMANIUM

Bridging group	End substituent	$K_c{}^a$	$K_{rc}{}^b$	$K_{rr}{}^c$ (liter mole^{-1})	Ref.
Me$_2$GeO	I	0.27	0.87	4.8	45
Me$_2$GeO	Br	0.043	0.87	4.8	45
Me$_2$GeO	Cl	0.02	0.87	4.8	45
Me$_2$GeS	I	0.09	50	—	42
Me$_2$GeS	Br	0.06	50	—	42
Me$_2$GeS	Cl	0.09	50	—	42
Cl$_2$GeNMe	Cl	0.14[d]	53[d] mole liter^{-1}	—	44

[a] K_c is defined by Eq. 7.17.
[b] K_{rc} is defined by Eq. 7.19 [34].
[c] $K_{rr} = [(\text{Me}_2\text{GeO})_4]^3/[(\text{Me}_2\text{GeO})_3]^4$ in moles of Me$_2$Ge.
[d] Branching chains as well as linear chains are possible in this system, so the constants are not directly comparable, but the amounts of branching groups are probably small [44]. The relevant equations have been derived by Matula *et al.* [34] but are difficult to apply since it is not certain what are the concentrations of the branching chain groups ClGe(NMe)$_3$ and Ge(NMe)$_4$ [44].

linked to germanium, which gradually are used up to form crosslinked amorphous polymers [44].

Equilibria for exchange of bridging sulfur or oxygen with halogen on dimethylgermanium moieties have been measured, including ring–ring, ring–chain, and chain–chain exchange [42, 45]. Equilibrium constants for families of compounds encompassed are given in Table 7.18. Exchanging substituents on dihalopolygermoxane chains (as neat liquids) have lifetimes of the order of $\frac{1}{10}$ sec at ambient temperatures, but exchange of substituents on rings is much slower. Kinetics have been compared for the system where bridging sulfur and halogen are exchanging on dimethylgermanium [42]: this exchange is slower than that for the germoxane system, but is nonetheless complete in a matter of hours at ambient temperatures.

$$
2\;
\begin{array}{c}
Me_2 \\
| \\
Ge \\
S^{\diagup}\;\;^{\diagdown}S \\
| \qquad | \\
Me_2Ge \quad GeMe_2 \\
^{\diagdown}O^{\diagup}
\end{array}
\;\rightleftharpoons\;
\begin{array}{c}
Me_2 \\
| \\
Ge \\
S^{\diagup}\;\;^{\diagdown}S \\
| \qquad | \\
Me_2Ge \quad GeMe_2 \\
^{\diagdown}S^{\diagup}
\end{array}
\;+\;
\begin{array}{c}
Me_2 \\
| \\
Ge \\
S^{\diagup}\;\;^{\diagdown}O \\
| \qquad | \\
Me_2Ge \quad GeMe_2 \\
^{\diagdown}O^{\diagup}
\end{array}
\qquad 7.23
$$

$$
2\;
\begin{array}{c}
Me_2 \\
| \\
Ge \\
S^{\diagup}\;\;^{\diagdown}O \\
| \qquad | \\
Me_2Ge \quad GeMe_2 \\
^{\diagdown}O^{\diagup}
\end{array}
\;\rightleftharpoons\;
\begin{array}{c}
Me_2 \\
| \\
Ge \\
O^{\diagup}\;\;^{\diagdown}O \\
| \qquad | \\
Me_2Ge \quad GeMe_2 \\
^{\diagdown}O^{\diagup}
\end{array}
\;+\;
\begin{array}{c}
Me_2 \\
| \\
Ge \\
S^{\diagup}\;\;^{\diagdown}S \\
| \qquad | \\
Me_2Ge \quad GeMe_2 \\
^{\diagdown}O^{\diagup}
\end{array}
\qquad 7.24
$$

The cyclic trimers $(Me_2GeO)_3$ and $(Me_2GeS)_3$ reacted to give trimers with mixed O and S backbones, as well as the equilibrium proportions of the two cyclic germoxanes $(Me_2GeO)_3$ and $(Me_2GeO)_4$ [46] already noted in Table 7.18 (K_{rr} value). Two further equilibria are needed to describe this system—Eqs. 7.23 and 7.24. Equilibrium constants for these equations were obtained as follows: for Eq. 7.23, 0.87 at 120° and 0.80 at 25°C; for Eq. 7.24, 0.59 at 120°C and 0.79 at 25°C [46].

III. Tin

Mixed alkyls of tin are formed in approximately random amounts in a system of two different stannic alkyls subject to exchange catalyzed by aluminum halide [47]. Recent studies of these systems employ gas chromatography as the separation method [48]. Two main approaches have been made to a quantitative evaluation of the scrambling of alkyl and halogen on tin [49, 50]. One used calorimetry to obtain reaction heats for liquid phase reactions at 25° and then the exact composition of the liquid in the calorimeter was determined by GLC analysis of its contents. It is necessary to obtain some quantitative information on reaction stoichiometry since there are *three* possible alkyltin halides and their relative amounts must be known. The column temperature used in the GLC analysis was 145°, but it was assumed that no alteration in composition of the mixture would ensue at this temperature [50]. The second approach was to examine the species in reaction mixtures of tetramethyltin and stannic chloride by ^{1}H NMR [49]. This has the advantage of an *in situ* method. The initial reaction 7.25 was found to occur almost immediately and prior to any other, with

$$Me_4Sn + SnCl_4 \rightleftharpoons MeSnCl_3 + Me_3SnCl \qquad 7.25$$

an activation energy of 15–21 kJ mole^{-1} [49] between 0 and 50°. For this reaction, equilibrium constant data obtained by NMR methods gave a value of $\delta \Delta G$ (see Eq. 7.12) of 96 kJ mole^{-1}, while the calorimetric data [50] give the reasonably close figure for ΔH of 92 kJ mole^{-1}. Equilibrium constants for Eqs. 7.26–7.27 were obtained from the NMR analyses. The

$$2MeSnCl_3 \rightleftharpoons Me_4Sn + Me_2SnCl_2, \qquad K = 3 \times 10^{-3} \qquad 7.26$$

$$2Me_2SnCl_2 \rightleftharpoons Me_3SnCl + MeSnCl_3, \qquad K = 1 \times 10^{-4} \qquad 7.27$$

$$2Me_3SnCl \rightleftharpoons Me_2SnCl_2 + SnCl_4, \qquad K = 7 \times 10^{-2} \qquad 7.28$$

experimental data obtained in the calorimetric work are given in Table 7.19. Values for more conventional redistribution equilibria are derived algebraically from these. Table 7.20 gives $\delta \Delta G$ values obtained algebraically from Eqs. 7.26–7.28 and ΔH values obtained algebraically [50] for the same stoichiometry, but the agreement is not good. Ethyl, vinyl, and butyl groups were also studied in the calorimetric study and the primary experimental data are shown in Table 7.19. The formation of R_2SnCl_2 was found to occur to a different extent in each system.

Redistribution of hydrogen with halogen is rapid and favors the mixed compound exclusively, quite unlike the related silicon systems [23, 51]. Alkyl and hydrogen redistribute exothermically but the reaction is closer to random than with alkyl and chloride [52] (Table 7.20). Alkyltin dichlorides with two different alkyl or aryl groups have been prepared by scrambling and GLC was used to follow the course of the reaction [53].

Bridging sulfur in trimeric dimethyltin sulfide $(Me_2SnS)_3$ exchanges with the neso molecules Me_2SnCl_2, Me_2SnBr_2, and Me_2SnI_2 to give polymeric systems for which the chain constant K_c (defined by Eq. 7.17) is near to the random figure (actual $K_c = 0.223 \pm 0.013$ in the iodine system) but the pre-exchange lifetimes of various species in these mixtures are very short on the

TABLE 7.19

HEATS OF LIQUID PHASE REACTIONS DETERMINED CALORIMETRICALLY[a] FOR SCRAMBLING
WHERE R IS Me, Vi, Et, AND Bu WITH Cl ON Sn

$$x R_4Sn + y SnCl_4 \rightarrow a RSnCl_3 + b R_2SnCl_2 + c R_3SnCl$$

R	x	y	a	b	c	$-\Delta H$ (kJ mole^{-1}) at 25°
Me	1	1.062	0.907	0.371	0.784	97 ± 4
Me	2.153	1	0.047	0.753	2.353	147 ± 6
Me	1	1.577	1.473	0.787	0.318	121 ± 6
Vi	1	1.01	0.14	1.75	0.12	88 ± 2
Vi	2.5	1.92	—	0.51	2.97	106 ± 2
Vi	1	1.92	1.84	1.08	—	97 ± 6
Et	1	0.9	0.90	0.05	1.00	96 ± 4
Et	1.09	1	0.91	0.09	1.09	94 ± 4
Et	1	1.33	1.33	0.33	0.67	106 ± 8
Bu	1	1	1	—	1	93 ± 12

[a] From Nash *et al.* [50].

TABLE 7.20

HEATS OF REACTION ΔH AND DEVIATION OF FREE ENERGY FROM IDEALITY $\delta \Delta G$
FOR EXCHANGE OF Me AND Cl OR H ON Sn[a]

Reaction	ΔH (kJ mole^{-1})	$\delta \Delta G$ (kJ mole^{-1})
$M_4 + M_0 \rightarrow M_1 + M_3$	-92	-96
$2M_4 + M_0 \rightarrow M_2 + 2M_3$	-143	-88
$0.75P_4$ (l) $+ 0.25P_0$ (g) $\rightarrow P_3$ (l)	-16.3	
$0.75B_4$ (l) $+ 0.25B_0$ (g) $\rightarrow B_3$ (l)	-15.9	

[a] ΔH for Me–Cl scrambling from data in Nash *et al.* [50]; $\delta \Delta G$ from Grant
and Van Wazer [49]; ΔH for Me–H scrambling from Stack *et al.* [52].

[b] M_x means Me_xSnCl_{4-x}; P_x means Pr_xSnH_{4-x}; B_x means Bu_xSnH_{4-x}.

NMR time scale and a full interpretation of the contents of the mixture is not at present possible [54].

Coupling of tin nuclei ^{117}Sn and ^{119}Sn (both with nuclear spin quantum number $I = \frac{1}{2}$) with the hydrogen nuclei present in alkyl groups R and R' in alkyldialkylaminotin compounds, $R_i Sn(NR_2')_{4-i}$, is observed in the form of satellites to the main ^{1}H resonance signals for the R and R' groups. The satellites to the R' groups on nitrogen disappear as the temperature of the system is raised, indicating some exchange process in which Sn—N, but not Sn—R bonds are broken [55].

IV. Scrambling between Group IV and Other Central Atoms

A. Mechanistic Studies

Several mixed reactions are pertinent to a study of mechanism in Group IV redistribution. Since the absolute configurations of a number of silanes have been established [56], the steric course of reactions at silicon can be examined. Most substitutions at silicon have been found to proceed with inversion of configuration at the central silicon—the transition state is conveniently thought of as a trigonal bipyramid (TBP) based on central silicon with the attacking and leaving groups axial. For several of the reactions we are about to discuss, the steric course is retention of configuration, which is rather more difficult to explain in terms of a TBP, since the attacking and leaving groups are required to be on the same side of the silicon, either both axial or one axial and one equatorial. Alkoxide exchange between lithium butoxide and optically pure methylphenyl-α-naphthylsilylmethoxide (I) proceeds with retention of configuration at silicon in benzene [57] (Eq. 7.29). The exchange of hydride and chloride between silicon and boron (Eq. 7.30)

$$\text{LiOBu} + \text{MePhNpSiOMe} \ \rightarrow \ \text{LiOMe} + \text{MePhNpSiOBu} \qquad 7.29$$
$$\text{I}$$

$$\text{BCl}_3 + \text{MePhNpSiH} \ \rightarrow \ \text{MePhNpSiCl} + \text{B}_2\text{H}_6 \qquad 7.30$$
$$\text{II}$$

has been found to proceed with retention of configuration at the silicon [58], likewise hydride exchange from lithium aluminum hydride to the silane (II) or its germanium analog [59]. In each of these reactions, a four-center transition state with exchanging groups on the same side of the silicon (or germanium) is consistent with the stereochemical result. Since the attacking and leaving groups are more polar than the alkyl and aryl substituents on silicon, it is difficult to reconcile such a transition state with current views on pseudorotation [60] or stereochemical nonrigidity [61] in the trigonal bipyramid. That retention occurs in nonpolar conditions for Eqs. 7.29 and 7.30 is consistent with little charge separation in a four-center mechanism. In solvents of high polarity (e.g., butanol) or with a more ionic butoxide (Na instead of Li), the steric course of Eq. 7.29 changes to inversion [57]. This is consistent with an S_N2 type of mechanism at silicon with a fully ionized alkoxide group as nucleophile and as leaving group, in axial positions on a trigonal-bipyramidal transition state.

The stereochemistry at silicon of a much more extensive series of exchanges between B and Si (Eq. 7.31 and 7.32) has now been examined

$$\text{MePhNpSiOR} + \text{BX}_3 \rightarrow \text{MePhNpSiX} \qquad \text{(R = Me, H, K, menthyl, } t\text{-Bu,} \quad 7.31$$
$$\text{X = F, Cl)}$$
$$\text{(Naphthyl could be replaced by benzhydryl and neopentyl)}$$

$$\text{MePhNpNRR}' + \text{BX}_3 \rightarrow \text{MePhNpSiX} \qquad \text{(R = H, R}' = \text{Bu;} \qquad 7.32$$
$$\text{R = R}' = \text{H;}$$
$$\text{R, R}' = \text{pyrrolidinyl)}$$

for a variety of substituents. The majority of these reactions proceed with inversion of configuration, except for 7.31 when R is Me, X is Cl, and 7.32 when R, R' is pyrrolidinyl and X is F. The proposed mechanism involves prior coordination of OR or NRR' to the boron halide, followed by a four center mechanism which leads to retention, or followed by an S_N2-type substitution of halogen for the leaving group, which leads to inversion [62].

The rate of exchange of tributyltin hydride with bis(dimethylamino)boron chloride has been shown to be independent of added free radicals, but to increase with solvent polarity, suggesting a polar mechanism [63].

The reaction between $GeCl_4$ and BBr_3 has been studied in a very large excess of $GeCl_4$, when it exhibits zero-order kinetics. The changes in concentration of various species with time are consistent with catalysis of the Cl/Br exchange by traces of HCl or HBr [64]. Reaction of tetramethyltin with the series $SnCl_4$, $GeCl_4$, $SiCl_4$ provides a typical rate comparison for Group IV atoms Sn $\gg$ Ge $\gg$ Si. While the Sn reaction occurs in minutes at room temperature, the silicon reaction is incomplete after many days at high temperatures [49].

B. Thermodynamic Studies

Exchange between MeSi$\equiv$ and MeGe$\equiv$ moieties has been treated by Moedritzer and Van Wazer using ^{1}H NMR techniques [65, 66]. The intersystem constants have been obtained at 120° and show that methylsilicon preferably binds the lighter halogen of a pair, and that the equilibria

TABLE 7.21

INTERSYSTEM EQUILIBRIUM CONSTANTS K_I FOR
SCRAMBLING OF X AND Y BETWEEN MeGe$\equiv$ AND
MeSi$\equiv$[a,b]

X	Y	K_I[c]
Cl	Br	$(9.7 \pm 4) \times 10^{-6}$
Br	I	$(4.2 \pm 0.9) \times 10^{-5}$
Cl	I	$(1.5 \pm 0.7) \times 10^{-12}$
Cl	OMe	$(1 \pm 10) \times 10^{14}$
Br	OMe	$(1 \pm 10) \times 10^{20}$
I	SMe	$(9.1 \pm 4.1) \times 10^{3}$
Br	SMe	(6.3 ± 1.5)
Cl	SMe	$(1.4 \pm 0.5) \times 10^{-7}$
Cl	NMe$_2$	$(3.5 \pm 3.0) \times 10^{12}$
Br	NMe$_2$	$(4.0 \pm 3.4) \times 10^{20}$
I	NMe$_2$	$(4.2 \pm 8.6) \times 10^{28}$

[a] From Moedritzer and Van Wazer [65, 66].
[b] Temperature, 120°C.
[c] $K_I = [\text{MeSiY}_3][\text{MeGeX}_3]/[\text{MeSiX}_3][\text{MeGeY}_3]$.

TABLE 7.22

INTERSYSTEM EQUILIBRIUM CONSTANTS K_I FOR
SCRAMBLING OF X AND Y BETWEEN Me$_2$Ge$=$ AND
Me$_2$Si$=$[a,b]

X	Y	K_I[c]
Cl	OMe	$(4.5 \pm 1.6) \times 10^{14}$
Br	OMe	$(9.2 \pm 5.4) \times 10^{12}$
I	OMe	$(1.14 \pm 0.21) \times 10^{10}$
Cl	SMe	$(1.02 \pm 0.27) \times 10^{-3}$
Br	SMe	(4.45 ± 0.51)
I	SMe	$(3.5 \pm 0.53) \times 10^{4}$
OMe	SMe	$(2.6 \pm 0.4) \times 10^{-4}$
Br	NC	$(2.63 \pm 1.0) \times 10^{2}$
Cl	NC	$(6.7 \pm 1) \times 10^{-2}$
I	Br	$(5.03 \pm 0.9) \times 10^{3}$
I	Cl	$(8.3 \pm 1.9) \times 10^{5}$
Cl	Br	$(2.9 \pm 1.2) \times 10^{-4}$

[a] From Van Wazer et al. [67] and Moedritzer and
Van Wazer [68].
[b] Temperature, 120°C.
[c] $K_I = [\text{Me}_2\text{SiY}_2][\text{Me}_2\text{GeX}_2]/[\text{Me}_2\text{SiX}_2][\text{Me}_2\text{GeY}_2]$.

are widely deviant from random. Of the pairs OMe–Cl, SMe–Cl, or NMe_2–Cl, the methylgermanium selectively attaches to the chloride and the methyl-silicon to the donor substituent (Table 7.21). The kinetics of the reactions are intermediate between those for Si–Si or Ge–Ge exchange of the same groups and a study has been made of the change in composition of the system with time starting from either side of the reaction for exchange of Cl and Br.

Exchange between $Me_2Ge=$ and $Me_2Si=$ is very similar: the intersystem equilibrium constants in Table 7.22 are similar in magnitude to those in Table 7.21, and the exchange rates are again intermediate between those for the all-silane or all-germane systems [67, 68]. The function $\delta \, \Delta G$ calculated from these intersystem K is sizable, and has been compared with the ΔH value calculated using Pauling's bond energy equation. The difference between $\delta \, \Delta G$ and ΔH should be slight since it pertains to entropy in theory, but the differences in fact are quite large and have been ascribed to π bonding [67, 68].

Exchange of bridging oxygen with halogen tends to go in the same direction as exchange of monofunctional oxygen (OMe). When halide and bridging oxygen are scrambled in mixtures of $Me_2Ge=$ and $Me_2Si=$ the germane tends to end up as the neso molecule and the silane as the polymer [69, 70]. The intersystem constant

$$K_I = [Si_{middles}][Ge_{neso}]/[Si_{neso}][Ge_{middles}]$$

was about 10^{10} for chloride [69] and for bromide [70]. Van Wazer and Moedritzer use another constant to describe the sequence of Ge and Si in the polymeric chains K_0

$$K_0 = [SiOSi][GeOGe]/[SiOGe]^2 = 0.35 \qquad (120°) \text{ for Cl}$$

In silicon-rich systems, pure siloxanes occur, but in germanium-rich compositions, germano-terminated polysiloxanes occur. Germanium in the middle of chains is thermodynamically disfavored [69, 70].

REFERENCES

1. I. R. Beattie, *Quart. Rev. Chem. Soc.* **17**, 382 (1963).
2. M. L. Delwaulle, *Bull. Soc. Chim. Fr.* **1951**, 485.
3. G. S. Forbes and H. H. Anderson, *J. Amer. Chem. Soc.* **66**, 931 (1944).
4. G. S. Forbes and H. H. Anderson, *J. Amer. Chem. Soc.* **67**, 1911 (1945).
5. M. L. Delwaulle, M. B. Buisset, and M. Delhaye, *J. Amer. Chem. Soc.* **74**, 5768 (1952).
6. J. R. Van Wazer and K. Moedritzer, *Inorg. Chem.* **3**, 268 (1964).
7. G. Calingaert and H. A. Beatty, *J. Amer. Chem. Soc.* **61**, 2748 (1939).

8. G. A. Russell, *J. Amer. Chem. Soc.* **81**, 4815, 4825, 4833 (1959).

9. J. A. Semlyen, G. R. Walker, R. E. Blofield, and C. S. G. Phillips, *J. Chem. Soc.* **1964**, 4948.

10. T. Flitcroft and H. A. Skinner, *J. Chem. Soc.* **1956**, 3335.

11. H. Weingarten and J. R. Van Wazer, *J. Amer. Chem. Soc.* **88**, 2700 (1966).

12. P. D. Zemany and F. P. Price, *J. Amer. Chem. Soc.* **70**, 4222 (1948).

13. K. Moedritzer and J. R. Van Wazer, *Inorg. Chem.* **5**, 547 (1966).

14. J. R. Van Wazer and K. Moedritzer, *J. Inorg. Nucl. Chem.* **26**, 737 (1964).

15. K. Moedritzer and J. R. Van Wazer, *Inorg. Chem.* **5**, 1254 (1966).

16. K. Moedritzer and J. R. Van Wazer, *J. Inorg. Nucl. Chem.* **29**, 1851 (1967).

17. K. Moedritzer and J. R. Van Wazer, *J. Organometal. Chem.* **12**, 69 (1968).

18. J. R. Van Wazer, K. Moedritzer, and L. C. D. Groenweghe, *J. Organometal. Chem.* **5**, 420 (1966).

19. K. Moedritzer and J. R. Van Wazer, *J. Organometal. Chem.* **6**, 242 (1966).

20. K. Moedritzer and J. R. Van Wazer, *Inorg. Chem.* **7**, 2105 (1968).

21. E. A. V. Ebsworth, A. G. Lee, and G. M. Sheldrick, *J. Chem. Soc. A* **1968**, 2294.

22. R. H. Baney and R. J. Shindorf, *J. Organometal. Chem.* **6**, 660 (1966).

23. D. R. Weyenberg, A. E. Bey, H. F. Stewart, and W. H. Atwell, *J. Organometal. Chem.* **6**, 586 (1966).

24. L. G. Mahone and D. R. Weyenburg, *J. Organometal. Chem.* **12**, 231 (1968).

25. K. Moedritzer and J. R. Van Wazer, *Inorg. Chem.* **6**, 93 (1967).

26. K. Moedritzer and J. R. Van Wazer, *Z. Anorg. Allg. Chem.* **345**, 35 (1966).

27. K. Moedritzer and J. R. Van Wazer, *J. Amer. Chem. Soc.* **86**, 802 (1964).

28. K. Moedritzer, J. R. Van Wazer, and C. H. Dungan, *J. Chem. Phys.* **42**, 2478 (1965).

29. J. R. Van Wazer and K. Moedritzer, *J. Chem. Phys.* **41**, 3122 (1964).

30. J. B. Carmichael and J. Heffel, *J. Phys. Chem.* **69**, 2213 (1965).

31. H. F. Stewart, *J. Organometal. Chem.* **10**, 229 (1967).

32. D. Grant, *J. Inorg. Nucl. Chem.* **29**, 69 (1967).

33. K. Moedritzer and J. R. Van Wazer, *J. Phys. Chem.* **70**, 2030 (1966).

34. D. W. Matula, L. C. D. Groenweghe, and J. R. Van Wazer, *J. Chem. Phys.* **41**, 3105 (1964).

35. G. M. Burch and J. R. Van Wazer, *J. Chem. Soc. A* **1966**, 586.

36. R. S. Bottei and L. J. Kuzma, *J. Inorg. Nucl. Chem.* **30**, 415 (1968).

37. R. S. Bottei and L. J. Kuzma, *J. Inorg. Nucl. Chem.* **30**, 2345 (1968).

38. K. Moedritzer and J. R. Van Wazer, *Inorg. Chem.* **5**, 547 (1966).

39. K. Moedritzer, J. R. Van Wazer, and R. E. Miller, *Inorg. Chem.* **7**, 1638 (1968).

40. K. Moedritzer and J. R. Van Wazer, *J. Organometal. Chem.* **13**, 145 (1968).

41. K. Moedritzer and J. R. Van Wazer, *J. Inorg. Nucl. Chem.* **29**, 1571 (1967).

42. K. Moedritzer and J. R. Van Wazer, *J. Amer. Chem. Soc.* **87**, 2360 (1965).

43. K. Moedritzer and J. R. Van Wazer, *J. Inorg. Nucl. Chem.* **28**, 957 (1966).

44. W. Eisenhuth and J. R. Van Wazer, *Inorg. Chem.* **7**, 1642 (1968).

45. K. Moedritzer and J. R. Van Wazer, *Inorg. Chem.* **4**, 1753 (1965).

46. K. Moedritzer and J. R. Van Wazer, *J. Amer. Chem. Soc.* **90**, 1708 (1968).

47. G. Calingaert and H. A. Beatty and H. R. Neal, *J. Amer. Chem. Soc.* **61**, 2755 (1939).

48. F. H. Pollard, G. Nickless, and P. C. Uden, *J. Chromatog.* **19**, 28 (1965).

49. D. Grant and J. R. Van Wazer, *J. Organometal. Chem.* **4**, 229 (1965).

50. G. A. Nash, H. A. Skinner, and W. F. Stack, *Trans. Faraday Soc.* **61**, 640 (1965).

51. A. K. Sawyer, J. E. Brown, and E. L. Hanson, *J. Organometal. Chem.* **3**, 464 (1965).

52. W. F. Stack, G. A. Nash, and H. A. Skinner, *Trans. Faraday Soc.* **61**, 2122 (1965).
53. H. G. Kuivila, R. Sommer, and D. C. Green, *J. Org. Chem.* **33**, 1119 (1968).
54. K. Moedritzer and J. R. Van Wazer, *Inorg. Chem.* **3**, 943 (1964).
55. E. W. Randall, Ch. H. Yoder, and J. J. Zuckerman, *J. Amer. Chem. Soc.* **89**, 3438 (1967).
56. L. H. Sommer, "Stereochemistry, Mechanism & Silicon." McGraw-Hill, New York, 1965.
57. L. H. Sommer and H. Fujimoto, *J. Amer. Chem. Soc.* **90**, 982 (1968).
58. C. J. Attridge, R. N. Haszeldine, and M. J. Newlands, *Chem. Commun.* **1966**, 911.
59. G. J. D. Peddle, J. M. Shafir, and S. G. McGeachin, *J. Organometal. Chem.* **15**, 505 (1968).
60. F. H. Westheimer, *Accounts Chem. Res.* **1**, 70 (1968).
61. E. L. Muetterties and R. A. Schunn, *Quart. Rev. Chem. Soc.* **20**, 245 (1966).
62. L. H. Sommer, J. D. Citron, and G. A. Parker, *J. Amer. Chem. Soc.* **91**, 4729 (1969).
63. H. C. Newsom and W. G. Woods, *Inorg. Chem.* **7**, 177 (1968).
64. M. J. Rand and F. Reizman, *Can. J. Chem.* **44**, 980 (1966).
65. K. Moedritzer and J. R. Van Wazer, *Inorg. Chem.* **5**, 547 (1966).
66. K. Moedritzer and J. R. Van Wazer, *Inorg. Chem. Acta* **1**, 407 (1967).
67. J. R. Van Wazer, K. Moedritzer, and L. C. D. Groenweghe, *J. Organometal. Chem.* **5**, 420 (1966).
68. K. Moedritzer and J. R. Van Wazer, *J. Inorg. Nucl. Chem.* **28**, 957 (1966).
69. K. Moedritzer and J. R. Van Wazer, *J. Polym. Sci. Part A* **6**, 547 (1968).
70. K. Moedritzer and J. R. Van Wazer, *J. Amer. Chem. Soc.* **90**, 47 (1968).

8

Group V

I. Phosphorus

A. *Polymers*

The chemistry of phosphorus has been especially important in the formulation of a theory of structural reorganization in inorganic chemistry. Stemming from the work of Van Wazer and others, came the theory that phosphates consisted of families of related compounds built up from a few kinds of grouping—the end (E), middle (M), and branching (B), groups together with the neso (N) or smallest molecule of the series—sometimes called the ortho molecule in phosphorus chemistry. The sequence of groups EE gives a

neso (N)   end (E)   middle (M)   branching (B)

pyrophosphate, the sequence EME is a trimer, long chains with sequences $E(M)_nE$ can be formed, and branching molecules can be introduced, for example, $B(ME)_3$. Van Wazer's theory of redistribution (structural reorganization) in polymeric families of compounds [1–3] seems to have developed from work on such phosphates. ^{31}P nuclear magnetic resonance

133

provided a major tool for the investigation of such condensed phosphate systems [4]. [31]P nuclear magnetic resonance signals show a wide range of chemical shifts so that signals are widely separated. Unfortunately coupling with other nuclei (including chemically distinct P,H in ester groupings, etc.) is strong, so that there will be fine structure complicating the spectra, and the effect of quadrupole broadening (e.g., in P–Cl or P–Br linkages) and viscosity broadening (quite serious in viscous polymers or hydrogen-bonded systems) can adversely affect the quality of resolution. Notwithstanding, ^{31}P NMR has enabled considerable precision in quantifying the thermodynamic description of phosphate systems [4, 5].

In a typical family of phosphorus compounds, the series N, E, M, and B, where X is Cl, was generated from a mixture of phosphoryl chloride (POCl$_3$) and the cage molecule P$_4$O$_{10}$ in various ratios. The equilibria between the building groups are given by Eqs. 8.1 and 8.2. (See also Chapter 7, Section IC.)

$$K_1 = [E][B]/[M]^2 \qquad\qquad 8.1$$

$$K_2 = [N][M]/[E]^2 \qquad\qquad 8.2$$

Weighted average values for the polyphosphate systems built up from E, B, M, and N where X is Cl are given in Table 8.1, with related values of

TABLE 8.1

SCRAMBLING OF BRIDGING —O— WITH MONOFUNCTIONAL SUBSTITUENTS
X ON $\geq$P=O SYSTEMS

End group (E)	K_2 (Eq. 8.2)	K_1 (Eq. 8.1)	Ref.
OP(O$_{1/2}$)(OH)$_2$	0.08	10^3	1
OP(O$_{1/2}$)Cl$_2$	0.64	0.422	5
OP(O$_{1/2}$)(NMe$_2$)$_2$	0.026	6×10^{-4}	6
OP(O$_{1/2}$)(OEt)$_2$	0.06	0.022	7
OP(O$_{1/2}$)(OMe)$_2$	0.058	0.054	8
OP(O$_{1/2}$)(OAc)$_2$	0.065	Not calculated	9
OP(O$_{1/2}$)(Me)(OMe)	0.05	—	10

K_1 and K_2 for the systems where X is OEt, OMe, NMe$_2$, OH, OAc, and one methylphosphonate system for which branched chains are not possible. The theoretical value of K_1 and K_2 for random sorting of the groups is 0.333.

The gel point in the chloro system should occur in theory at R = [Cl]/[P] = 1.22 which is found in practice [5]. The existence of rings in such mixtures complicates the equilibria (Eq. 7.19, see Chapter 7, Section IC).

The systems mentioned in Table 8.1 can be built up in several different ways and not necessarily by scrambling. The methylphosphonates, for instance, can be made from condensation of $MePOCl_2$ with $MePO(OMe)_2$ [10] by elimination of methyl chloride to give bridging oxygen groups. ^1H nuclear magnetic resonance has been used where applicable in the studies of Table 8.1. There are no cyclic molecules [10] in methoxy-terminated polyphosphonates, but the constants K are not totally consistent unless allowance is made for a reorganizational heat order. This could not be fully evaluated because of the complexity of the system but appeared to be about 2 [2, 10]. The variation with time of the concentration of various species is discussed [10].

Redistribution is involved in the build up of cyclic and linear phosphonitrilic polymers. Conversion of trimer $(PNCl_2)_3$ or tetramer $(PNCl_2)_4$ to the high polymer $(PNCl_2)_n$ seems to occur in an ionic chain mechanism. The polymerization is first order in trimer concentration with an activation energy of 102 ± 4 kJ mole^{-1} and the enthalpy of polymerization observed thermochemically was 5.8 kJ mole^{-1} [11, 12].

B. Monomeric Systems

One has to consider scrambling of monofunctional substituents on trigonal P⟨ tetrahedral P⟨ (as in O=P or S=P⟨) and trigonal-bipyramidal phosphorus. Some data are given in Table 8.2 corresponding to Eqs. 8.3

$$2MX_2Y \rightleftharpoons MXY_2 + MX_3 \qquad 8.3$$

$$2MXY_2 \rightleftharpoons MY_3 + MX_2Y \qquad 8.4$$

and 8.4. All the constants in Table 8.2 have values close to random, except those in the first two rows. Scrambling on trigonal P of NEt_2 and Cl is nonrandom, in fact the K values could not readily be computed since the amounts of unscrambled species were so small. Enthalpies of formation for $P(NEt_2)_xCl_{3-x}$ were about 38 kJ mole^{-1} [17].

Intersystem scrambling on phosphorus as P⟨, O=P⟨, S=P⟨ is described in Table 8.3 together with some other equilibria. There is little departure from random in the intersystem scrambling but chlorine bonds selectively to O=P⟨ and bromine to S=P⟨. Most of the data in Table 8.3 were obtained using NMR spectra, but in one case Raman spectroscopy

TABLE 8.2

REDISTRIBUTION OF SUBSTITUENTS X AND Y ATTACHED TO PHOSPHORUS

M	X	Y	K (Eq. 8.3)	K (Eq. 8.4)	Conditions[a]	Ref.
P	OPh	N(Et)$_2$	0.0015	0.0016		13
P	Cl	OEt	0.0015	0.0014		13
P	Cl	SBu	0.157 ± 0.05	0.302 ± 0.008	2 days, 120°C	13
P	SMe	SBu	0.335 ± 0.05	0.322 ± 0.004	<3 days, 120°C	13
P=O	OMe	OEt	0.224	0.348	<7 days, 200°C	13
P	OMe	OEt	0.394 ± 0.019	0.367 ± 0.02	10–15 hr, 120°C	13
P	Br	OPh	0.38	0.12	12 hr, 180°C	14
P	Cl	OPh	0.13	0.13	12 hr, 180°C	14
P=S	Cl	Br	0.297 ± 0.012	0.419 ± 0.019	<1 week, 130°C	16
P=O	Cl	Br	0.406 ± 0.013	0.341	<1 week, 130°C	15
P	Cl	Br	0.2	0.7	<15 min, 25°C	14

[a] Time required for equilibrium starting from MX$_3$ and MY$_3$.

TABLE 8.3

SOME SCRAMBLING REACTIONS ON PHOSPHORUS, INCLUDING INTERSYSTEM SCRAMBLING

K value	Concentration ratio	Temp (°C)	Ref.
0.286	[HPO(OCH$_3$)$_2$][HPO(OEt)$_2$]/[HPO(OMe)(OEt)]^2	150	13
6.7	[POCl$_3$][PSBr$_3$]/[PSCl$_3$][POBr$_3$]	230	15
3.1	[POCl$_2$Br][PSBr$_3$]/[PSCl$_2$Br][POBr$_3$]	230	15
3.4	[POCl$_3$][PSClBr$_2$]/[POClBr$_2$][PSCl$_3$]	230	15
2.0	[POClBr$_2$][PSBr$_3$]/[PSClBr$_2$][POBr$_3$]	230	15
1.6	[POCl$_2$Br][PSClBr$_2$]/[POClBr$_2$][PSCl$_2$Br]	230	15
2.2	[POCl$_3$][PSCl$_2$Br]/[PSCl$_3$][POCl$_2$Br]	230	15
3.6	[POCl$_3$][ClCH$_2$PSCl$_2$]/[ClCH$_2$POCl$_2$][PSCl$_3$]	230	15
1.75	[POCl$_3$][PhPSCl$_2$]/[PSCl$_3$][PhPOCl$_2$]	230	15
1.3	[MePSCl$_2$][POCl$_3$]/[MePOCl$_2$][PSCl$_3$]	230	15
3.7	[PFClBr]^2/[PFCl$_2$][PFBr$_2$]	Ambient	18
1500	[PClBr(OPh)]^3/[PCl$_3$][PBr$_3$][P(OPh)$_3$]	180	14
5	[PBr$_2$Cl][POBrCl$_2$]/[POBr$_2$Cl][PBrCl$_2$]	200	19

was used (the mixed PFClBr). From vibrational spectra, molar heats, entropies and enthalpies from 200–1000°K have been calculated by the rigid rotator, harmonic oscillator approximation [20] (Chapter 2, Section I), for all the mixed compounds MX_2Y, $M = PO$ and PS; X, $Y = F$, Cl, and Br.

The enthalpy of formation of alkylaminodifluorophosphines is such that no redistribution is expected. However prolonged heating of an excess of R_2NPF_2 with a platinum halide ($PtCl_2$, $PtBr_2$) causes formation of PF_3 and the $(R_2N)_2PF$ *as its complex* with platinum halide. This is surely attributable to strong complexing of the metal—it acts as a "sink" for $(R_2N)_2PF$ forcing reaction in this direction. The bromide was more effective than the chloride and/or bromide in reactions 8.5–8.10.

The unusual series $POF_x(OSO_2F)_{3-x}$ where $x = 0-3$ has been identified by ^{19}F NMR in the decomposition mixture from $PO(OSO_2F)_3$. This evolves SO_3 to form the other members of the series. The equilibrium position is not known, but would be of interest for linear free energy correlations, since this is such an unusual ligand (OSO_2F) [22].

A number of possible halogen exchanges in mixtures of phosphorus, thiophosphoryl, and phosphoryl halides can be identified and some comments on them are warranted here. The exchanges are shown for chloride and/or bromide in reactions 8.5–8.10.

$$PCl_3 + PBr_3 \rightleftharpoons PCl_2Br + PClBr_2 \qquad\qquad 8.5$$

$$PCl_3 + {}^*POBr_3 \rightleftharpoons {}^*POCl_3 + PBr_3 \qquad\qquad 8.6$$
$$\text{1 week at } 200° \text{ [19]}$$

$$POCl_3 + POBr_3 \rightleftharpoons POCl_2Br + POClBr_2 \qquad\qquad 8.7$$
$$\text{1 week at } 130° \text{ [16]}$$

$$POCl_3 + {}^*PSBr_3 \rightleftharpoons POBr_3 + {}^*PSCl_3 \qquad\qquad 8.8$$
$$\text{hr at } 230° \text{ [15]}$$

$$RPOCl_2 + PSCl_3 \rightleftharpoons RPSCl_2 + POCl_3 \qquad\qquad 8.9$$
$$\text{hr at } 230° \text{ [15]}$$

$$PCl_3 + {}^*POCl_3 \rightleftharpoons {}^*PCl_3 + POCl_3 \qquad\qquad 8.10$$
$$\text{catalyzed by moisture or radiation [23]}$$

The asterisks are intended to show which groups exchange. Sometimes an alkyl or aryl (R) on phosphorus has been used as a label, under conditions

where the P—R bond will not scramble. The experimental conditions used
to bring each reaction to equilibrium are noted in brackets after the appropriate equation or in Table 8.2. The equilibrium positions for reactions
8.5 and 8.7 are given in Table 8.2, and those for reactions 8.6, 8.8, and
8.9 are given in Table 8.3. Reaction 8.10 represents a change in oxidation
state of phosphorus and a study of its mechanism was made using the ^{32}P
isotope as a label. Traces of moisture cause the reaction to occur, but under
uv or γ-irradiation, it took place at 25 or 50° [23]. Two competing mechanisms—one first order in phosphoryl chloride and phosphorus trichloride,
and one first order in phosphoryl chloride only—were compatible with the
observed rate dependence (Eq. 8.11). The first-order mechanism proposed is

$$\text{amount of } {}^{32}\text{P exchange} \propto [\text{POCl}_3] + k[\text{POCl}_3][\text{PCl}_3] \qquad 8.11$$

given by 8.12, and the second-order exchange proposed is given by Eq. 8.13.

$$^{32}\text{POCl}_3 \xrightarrow{\textit{hv, slow}} {}^{32}\text{PCl}_3 + \text{O}$$

$$\text{O} + \text{PCl}_3 \xrightarrow{\textit{fast}} \text{POCl}_3 \qquad\qquad 8.12$$

$$^{32}\text{POCl}_3 + \text{PCl}_3 \xrightarrow[\textit{hv}]{\textit{slow}} [\text{Cl}_3{}^{32}\text{P} \cdots \text{O} \cdots \text{PCl}_3]$$

$$\longrightarrow \text{Cl}_3{}^{32}\text{P} + \text{POCl}_3 \qquad 8.13$$

An approximate rate order for exchanges like Eq. 8.5 of F < Cl < Br < I
[24] has been reported.

II. Arsenic

Thermodynamic behavior in arsenic [As(III)] systems is comparable to
that in phosphorus [P(III)] systems (compare Si, Ge, and Sn) but reactions
are always faster in arsenic chemistry (again, compare Si, Ge, and Sn).

Exchange of chlorine and bromine on arsenic is random [25, 26] but the
mixed halide is not thermodynamically favored in exchange of fluorine with
chlorine [27]. The chloride–bromide exchange was studied by Raman
spectra, which indicated random exchange, and also by exchange of the
radioactive tracer ^{76}As in mixtures. Only pure arsenic chloride or bromide
could be obtained by fractionation, so the disproportionation of the mixed
halide must be rapid at the boiling point of the chloride (132°). No lines

which could be attributed to a mixed chloride-fluoride were observed in Raman spectra of mixtures of arsenic trifluoride and trichloride [26]. It seems that the concentration of mixed halide formed is small, which accounts for the Raman observation. More sensitive techniques have detected small amounts of mixed halides. In the mass spectra of mixtures of arsenic trifluoride and trichloride, the parent ions $AsCl_2F^+$ and $AsClF_2^+$ appear, with the correct m/e ratio and isotopic peaks [27]. Liquid-phase ^{19}F NMR showed three signals, one corresponding to AsF_3, but the others cannot be firmly assigned because there is the possibility of self-ionization of the arsenic fluoride to give ionic species. From the areas of the signals tentatively assigned as $AsCl_2F$ and $AsClF_2$ the constants for the equilibria 8.14 and 8.15 can be obtained. Corresponding equilibrium constants can be

$$2AsF_3 + AsCl_3 \ \rightleftharpoons \ 3AsF_2Cl, \qquad K = 4.9 \pm 1.4 \times 10^{-3} \qquad\qquad 8.14$$

$$2AsCl_3 + AsF_3 \ \rightleftharpoons \ 3AsFCl_2, \qquad K = 4.6 \pm 1.4 \times 10^{-3} \qquad\qquad 8.15$$

derived from the peak heights of the parent ions in the mass spectra of the mixtures, if it is assumed that the sensitivities of the species on either side of Eqs. 8.14 and 8.15 cancel. The apparent K values obtained in this way [27] were $(8.1 \pm 0.9) \times 10^{-3}$ (Eq. 8.14) and $(5.9 \pm 0.3) \times 10^{-3}$ (Eq. 8.15). (See Chapter 2, Section III.)

The exchanges of halogen with alkoxide, dialkylamino, or phenyl groups on arsenic are all strongly exothermic. Whereas the alkoxide and dialkyl-amino reactions are all extremely rapid on the ^1H NMR time scale [28], with lifetimes before exchange of about 0.01 sec (NMe_2 with F, Cl, or Br) and activation energies about 8–20 kJ mole^{-1} (NMe_2), the exchange of phenyl and chloride (Eqs. 8.16 and 8.17, Table 8.4) requires temperatures of 250–300° and has an activation energy (Eq. 8.17, k_2) of 157 kJ mole^{-1} and (Eq. 8.17, k_{-2}) of 143 kJ mole^{-1}, with corresponding activation entropies of -12 and -18 J °K^{-1} [29]. The second-order rate constants for the two reactions, Eqs. 8.16 and 8.17, are shown in Table 8.4. Equilibrium constants for a number of these exchanges on arsenic are given in Table 8.5.

$$2PhAsCl_2 \ \underset{k_{-1}}{\overset{k_1}{\rightleftharpoons}} \ Ph_2AsCl + AsCl_3 \qquad\qquad 8.16$$

$$2Ph_2AsCl \ \underset{k_{-2}}{\overset{k_2}{\rightleftharpoons}} \ Ph_3As + PhAsCl_2 \qquad\qquad 8.17$$

Pentavalent organoarsenic dihalides exchange halogen in solution [30]. Redistribution reactions of Me_3AsX_2 and $(PhCH_2)_3AsX_2$ where X is halogen, have been investigated with ^1H and ^{19}F NMR spectra and signal areas have been used where practicable to compute equilibrium constants at low temperatures, where spectra are resolved (Table 8.6). Exchanges between F–Cl and Cl–Br were too rapid at $-60°$ for the individual resonances of the reaction components to be observed [30].

TABLE 8.4

SECOND-ORDER RATE CONSTANTS FOR EQS. 8.16
AND 8.17

Temp (°C)	$k \times 10^3$ cm^3 mole^{-1} sec^{-1}	
304	$k_2 = 22.7$	$k_{-2} = 284$
256	$k_1 = 3.22$	$k_{-1} = 56.6$
252	$k_2 = 0.89$	$k_{-2} = 15.1$

TABLE 8.5

REDISTRIBUTION OF SUBSTITUENTS X AND Y ATTACHED TO ARSENIC [As(III)]^a

Method	X	Y	K (Eq. 8.3)	K (Eq. 8.4)	Temp (°C)	Ref.
Raman	Cl	Br	Near 0.33	Near 0.33	Ambient	26
Mass spectroscopy, NMR	Cl	F	$0.17–0.22 \times 10^3$	$0.12–0.20 \times 10^3$	Ambient	27
Chemical analysis	Ph	Cl	5.7×10^{-2}		256	29
Chemical analysis	Ph	Cl		5.9×10^{-2}	252	29
Chemical analysis	Ph	Cl	7.0×10^{-2}	8.2×10^{-2}	304	29
NMR	OMe	Br	0.016	0.009	37	28
NMR	OMe	Cl	7.3×10^{-4}	8.6×10^{-2}	37	28
NMR	OMe	F	1.4×10^{-2}	8.0×10^{-2}	28	28
NMR	OMe	F	1.9×10^{-2}	6.8×10^{-2}	37	28
NMR	NMe$_2$	OMe	9.5×10^{-2}	0.314	37	28
NMR	NMe$_2$	F	$\sim 10^{-8}$	1.6×10^{-2}	37	28

^{a} M = As(III) throughout.

TABLE 8.6

Equilibrium Constants K for the Scrambling
of X and Y on R_3As Determined by NMR

$$R_3AsF_2 + R_3AsCl_2 \overset{K}{\rightleftharpoons} 2R_3AsFCl^a$$

R	K	Temp (°C)
Me	5.4 ± 0.6	-53
PhCH$_2$	10 ± 1	-47

^{a} Even at $-60°$ individual NMR signals are not
observed for the F–Br and Cl–Br exchanges.

The cage compounds arsenic trioxide, As_4O_6, and tetraarsenic hexa-
methylimide, $As_4(NMe)_6$, have oxygen or nitrogen bridges which can be
made to scramble with arsenic compounds AsX_3 to produce families of
compounds based on the units E, M, and B with the neso compound N
[31–33]:

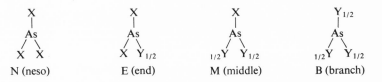

X is a monofunctional substituent (F, Cl, OMe, and NMe$_2$) and Y is a
difunctional substituent capable of acting as a bridge between arsenic atoms
(O, NMe, and S). The composition of each entire family for a specific X and
Y can be expressed simply by the equilibrium constants for Eqs. 8.1 and
8.2, assuming that no rings are formed. Various mixtures of As_4O_6 and
$As_4(NMe)_6$ with AsX_3 compounds have been examined using ^1H or ^{19}F
NMR and the occurrence of structural reorganization to give linear and
branching chains based on E, M, and B has been established. Unfortunately,
structural reorganization is sufficiently fast that individual species cannot be
isolated from the mixtures. The identity of the various NMR signals in the
mixtures can thus not be established beyond doubt by comparison with the
authentic compounds. Assumptions have to be made about various possible
assignments of NMR signals to different chemical environments, and then

the signal areas are processed by computer to provide the most self-consistent set of values for K_1 and K_2. It is necessary to work in the compositions where few rings are expected, or to assume the absence of rings since no means exist at present of relating NMR signals observed to ring structures in these arsenic families.

The most satisfactory reduction of experimental data has been obtained for the exchange of F with bridging O, which seems to be close to random. The evaluated constants for Eqs. 8.1 and 8.2 are shown in Table 8.7 for various arsenic systems. The activation energy for exchange of ^{19}F through structural reorganization in the F–O system appears to be the same as that for viscous flow. The data are shown in Table 8.8, along with ^{19}F exchange

TABLE 8.7

EQUILIBRIUM CONSTANTS FOR FORMATION OF LINEAR AND BRANCHING CHAINS IN POLYMERS OF ARSENIC ACCORDING TO EQS. 8.1 AND 8.2

Bridging group	Terminal	K_2 (Eq. 8.2)	K_1 (Eq. 8.1)	Ref.
O	F	0.33 ± 0.06	0.31 ± 0.02	31
O	OMe	0.36 ± 0.2	0.29 ± 0.15	33
O	NMe$_2$	~ 0.2	~ 4	33
NMe	F	$0.09–0.11$	$0.1–0.001$	32

TABLE 8.8

ACTIVATION PARAMETERS FOR ^{19}F OR ^1H EXCHANGE OBTAINED BY NMR METHODS, AND ACTIVATION ENERGIES FOR VISCOUS FLOW IN CERTAIN FAMILIES OF POLYARSENOUS OXIDES[a]

Bridging group Y	Terminal X	$E^{\ddagger}_{exchange}$ (kJ mole^{-1})	$E^{\ddagger}_{viscous\ flow}$ (kJ mole^{-1})	$R = [X]/[As]$
O	OMe	8–12		1.99
O	OMe	17		1.56
O	F	18	17.2	2.0
O	F	39.8	38	1.25

[a] From Van Wazer *et al.* [31, 33].

data for other systems. This suggests that viscous flow occurs by shearing of As—F bonds in the redistribution process, i.e., that viscous flow comes about through an actual chemical reaction [31]. The theoretical gel points were not observed in practice in these studies. The gel-point concept may of course be inapplicable in the face of rapid redistribution processes (which enable viscous flow) or it may be that rings occur. Exchange of fluorine with bridging NMe is very far from random (cf. Table 8.5) and the equilibrium constant values (Table 8.7) were dependent on the ratio [F]/[As]. Complete assignment of all NMR peaks could not be made and the K values obtained are not very consistent. Rapid exchange also occurs in this family [32].

Arsenic trioxide scrambles on heating with trimethyl arsenite [As(OMe)$_3$] or tris(dimethylamino)arsine [32]. The elimination of trimethylamine suggests the latter exchange is more than just a redistribution. The compound arsenic trisulfide (As$_2$S$_3$) probably scrambles with tris(dimethylamino) arsine, but again the reaction is complicated since realgar (AsS) is precipitated from the reaction mixture [33].

III.   Nitrogen, Antimony, and Bismuth

A.   *Nitrogen*

Mixed halides of nitrogen are known, but no scrambling reactions thereof. The scrambling of hydrogen and deuterium on nitrogen, Eq. 8.18, is probably as close to ideal random scrambling as possible [34]. The ^{15}N NMR

$$NH_3 + ND_3 \rightleftharpoons NH_2D + NHD_2 \qquad 8.18$$

spectra of ^{15}ND$_3$, ^{15}ND$_2$H, ^{15}NDH$_2$, and ^{15}NH$_3$ have been studied [34] and show near-linear chemical shifts as do the ^{1}H spectra of the mixture in Eq. 8.18. Table 8.9 contains a list of chemical shifts and coupling constants. An acid–base mechanism is possible for this exchange, although whether by separation of charged species or by a four-center "push–pull" mechanism is not known.

B.   *Antimony*

A mixed halide of antimony (SbBrI$_2$) has been isolated as a solid [35]. Exchange of radioactive tracer ^{122}Sb between antimony trichloride and

TABLE 8.9

CHEMICAL SHIFTS AND COUPLING CONSTANTS OBSERVED IN ^{15}N AND ^1H NMR SPECTRA
OF $^{15}ND_3$–$^{15}NH_3$ MIXTURES[a]

	$^{15}NH_3$	$^{15}NH_2D$	$^{15}NHD_2$	$^{15}ND_3$
^{15}N chemical shift (ppm)	0	0.68 ± 0.03	1.29 ± 0.03	1.96 ± 0.03
^1H chemical shift (ppm)	0	0.029 ± 0.002	0.053 ± 0.003	—
J ^{15}N–^1H (Hz)	61.8 ± 0.5	61.8 ± 0.5	61.8 ± 0.5	—
J ^{15}N–^2D (Hz)	—	9.45 ± 0.2	9.45 ± 0.2	9.45 ± 0.2
J ^1H–^2D (Hz)	—	1.54 ± 0.12	1.54 ± 0.12	—

[a] From Litchman [34].

TABLE 8.10

EQUILIBRIUM CONSTANTS K FOR THE SCRAMBLING OF X AND Y ON R_3Sb⚌

$$R_3SbX_2 + R_3SbY_2 \overset{K}{\rightleftharpoons} 2R_3SbXY \qquad (K_{random} = 4)$$

R	X	Y	K (60°C)	K (35°C)	K (0°C)	K (−37°C)
Ph	F	Cl		9.0 ± 1.2		
Ph	F	Br		4.0 ± 0.7		
Ph	F	I		1.0 ± 0.3		
PhCH$_2$	F	Cl				6.4 ± 0.5
Me	F	Cl	4.20 ± 0.10	3.90 ± 0.15		
Me	F	Br	3.61 ± 0.19	3.24 ± 0.14	2.81 ± 0.09	
Me[a]	F	I	1.68 ± 0.13	1.50 ± 0.06	1.09 ± 0.10	
Me	Cl	Br	3.65 ± 0.13	3.50 ± 0.15	3.28 ± 0.20	
Me	Cl	I	2.85 ± 0.08	2.25 ± 0.10	2.05 ± 0.07	
Me	Br	I	3.30 ± 0.16	3.02 ± 0.10	2.93 ± 0.03	

[a] For this reaction $\delta \Delta G = 2.5$ kJ mole^{-1}, $\Delta H = 5.5$ kJ mole^{-1}.

antimony tribromide [36] has been observed. Halogen exchange between
pentavalent organoantimony halides has been investigated by NMR
(^1H or ^{19}F) to give the equilibrium constants

$$K = [R_3SbXY]^2/[R_3SbX_2][R_3SbY_2]$$

which are shown in Table 8.10.

The exchange of groups between antimony trimethyl and antimony trichloride was found to be first order in each reagent in dimethylformamide solvent (Eq. 8.19). A typical second-order rate constant at $72°$ was $k_2 = 0.67 \pm 0.5 \times 10^{-4}$ liter mole^{-1} sec^{-1}. The activation energy was found to be

$$SbMe_3 + SbCl_3 \rightleftharpoons SbCl_2Me + SbClMe_2 \qquad 8.19$$

68.5 kJ mole^{-1} and the entropy of activation was -105 J $°K^{-1}$. The large negative entropy is in keeping with a four-center transition state [37].

C. Bismuth

Mixed chloride–bromide complexes have been investigated by uv spectroscopy [38]. A spectrophotometric method for determining equilibrium constants in systems $MA_3 + MB_3$, where there is no ionization of either to give B^-, A^-, etc., has been developed [39]. A swamping excess of one reagent pushes the equilibrium over to form just one of the possible mixed compounds and excess of starting reagent, so that the characteristic spectrum of the mixed compound can be determined. For BiI_3 and $BiCl_3$ in dioxane, it was possible to determine the constants

$$K_{21} = [BiI_2Cl]^2/[BiICl_2][BI_3] = 0.16 \pm 0.05$$

$$K_{12} = [BiICl_2]^2/[BI_2Cl][BiCl_3] = 3.6 \pm 0.2$$

at $25°$. The statistical value is 4. Since K_{21} is so small, it is possible that dioxane is coordinating to the bismuth [39].

REFERENCES

1. J. R. Van Wazer, "Phosphorus and Its Compounds," Vol. I. Wiley (Interscience), New York, 1958.
2. D. W. Matula, L. C. D. Groenweghe, and J. R. Van Wazer, *J. Chem. Phys.* **41**, 3105 (1964).
3. J. R. Van Wazer, *Amer. Sci.* **50**, 450 (1962).
4. M. M. Crutchfield, C. H. Dungan, J. H. Letcher, V. Mark, and J. R. Van Wazer, "Topics in Phosphorus Chemistry," Vol. 5. Wiley (Interscience), New York, 1967.
5. L. C. D. Groenweghe, J. H. Payne, and J. R. Van Wazer, *J. Amer. Chem. Soc.* **82**, 5305 (1960).
6. E. Schwarzmann and J. R. Van Wazer, *J. Amer. Chem. Soc.* **82**, 6009 (1960).
7. E. Schwarzmann and J. R. Van Wazer, *J. Amer. Chem. Soc.* **83**, 365 (1961).
8. J. R. Van Wazer and S. Norval, *J. Amer. Chem. Soc.* **88**, 4415 (1966).
9. C. Doremieux-Morin, J. M. Verdier, and R. Vincent, *Bull. Soc. Chim. Fr.* **1967**, 1628.
10. D. Grant, J. R. Van Wazer, and C. H. Dungan, *J. Polym. Sci. Part A* **1**, 57 (1967).

11. J. O. Konecny and C. M. Douglas, *J. Polym. Sci.* **36**, 195 (1959).
12. J. O. Konecny, C. M. Douglas, and M. Y. Gray, *J. Polym. Sci.* **42**, 383 (1960).
13. K. Moedritzer, G. M. Burch, J. R. Van Wazer, and H. K. Hofmeister, *Inorg. Chem.* **2**, 1152 (1963).
14. E. Fluck, J. R. Van Wazer, and L. C. D. Groenweghe, *J. Amer. Chem. Soc.* **81**, 6363 (1959).
15. L. C. D. Groenweghe and J. H. Payne, *J. Amer. Chem. Soc.* **83**, 1811 (1961).
16. L. C. D. Groenweghe and J. H. Payne, *J. Amer. Chem. Soc.* **81**, 6357 (1959).
17. J. R. Van Wazer and L. Maier, *J. Amer. Chem. Soc.* **86**, 811 (1964).
18. M.-L. Delwaulle, M. Cras, and P. M. Bridoux, *Bull. Soc. Chim. Fr.* **1960**, 786.
19. E. Schwarzmann and J. R. Van Wazer, *J. Amer. Chem. Soc.* **81**, 6366 (1959).
20. H. G. Horn and A. Mueller, *Z. Naturforsch.* **21A**, 431 (1966).
21. J. F. Nixon and M. D. Sexton, *Chem. Commun.* **1969**, 827.
22. D. D. des Marteau and G. H. Cady, *Inorg. Chem.* **5**, 1829 (1966).
23. L. F. Grantham and H. C. Moser, *J. Phys. Chem.* **66**, 863 (1962).
24. A. H. Cowley and S. T. Cohen, *Inorg. Chem.* **4**, 1221 (1965).
25. R. Muxart and R. Daudel, *J. Chim. Phys.* **47**, 610 (1950).
26. F. Francois and M.-L. Delwaulle, *J. Chim. Phys.* **46**, 80 (1949).
27. J. K. Ruff and G. Paulett, *Inorg. Chem.* **3**, 998 (1964).
28. K. Moedritzer and J. R. Van Wazer, *Inorg. Chem.* **3**, 139 (1964).
29. A. G. Evans and E. Warhurst, *Trans. Faraday Soc.* **44**, 189 (1948).
30. C. G. Moreland, M. H. O'Brien, C. E. Douthit, and G. G. Long, *Inorg. Chem.* **4**, 834 (1968).
31. J. R. Van Wazer, K. Moedritzer, and D. W. Matula, *J. Amer. Chem. Soc.* **86**, 807 (1964).
32. M. D. Rausch, J. R. Van Wazer, and K. Moedritzer, *J. Amer. Chem. Soc.* **86**, 814 (1964).
33. K. Moedritzer and J. R. Van Wazer, *Inorg. Chem.* **4**, 893 (1965).
34. W. M. Litchman, M. Alei, Jr., and A. E. Florin, *J. Chem. Phys.* **50**, 1897 (1969).
35. R. E. D. Clark, *J. Chem. Soc.* **1930**, 2737.
36. R. Muxart and R. Daudel, *J. Chim. Phys.* **47**, 610 (1950).
37. H. Weingarten and J. R. Van Wazer, *J. Amer. Chem. Soc.* **88**, 2700 (1966).
38. L. Newman and D. N. Hume, *J. Amer. Chem. Soc.* **79**, 4581 (1957).
39. F. Gaizer and M. T. Beck, *J. Inorg. Nucl. Chem.* **28**, 503 (1966).

9

Group VI

I. Sulfur

Scrambling reactions based on a central sulfur atom are known for sulfur in the II, IV, and VI oxidation states.

A. Sulfides and Sulfanes: Oxidation State II

Exchange reactions possible in polysulfide systems with alkyl or other end groups are exchange of end groups without change in chain length, exemplified in Eq. 9.1, and the slower disproportionation of sulfur chains,

$$Et_2S_3 + nPr_2S_3 \rightleftharpoons 2EtnPrS_3 \qquad 9.1$$

exemplified for one of the same reaction components in Eq. 9.2, which can

$$2Et_2S_3 \rightleftharpoons Et_2S_2 + Et_2S_4 \qquad 9.2$$

occur concurrently with exchange of the first type but appears to be slower. The mechanisms of these reactions are of importance in connection with the sulfur vulcanization of rubber and the role of disulfide interchange in deformation of keratin fibers [1–3]. Reactions of the first type have been shown to have equilibrium constants close to the statistically expected value of 4 where there are no steric complications in the alkyl end groups [3].

Haraldson *et al.* [3] investigated some dialkyl disulfide exchange reactions of the first kind thermodynamically at two temperatures, using gas chromatography detection techniques: the results are shown in Table 9.1.

TABLE 9.1

THERMODYNAMIC DATA FOR THE REACTION $Et_2S_2 + R_2S_2 \rightleftharpoons 2EtRS_2$[a]

R	K_{298} (± 0.2)[b]	ΔF_{298} (± 0.015) (kJ mole^{-1})	ΔH_{315} (± 0.15) (kJ mole^{-1})	ΔS (J $^{\circ}$K^{-1})
Me	5.1	-4.27	0.0	14.3
i-Pr	4.15	-3.52	—	11.8
t-Bu	24	-7.94	0.0	26.6

[a] From Haraldson *et al.* [3].
[b] Statistical value of K is 4.

The reactions can be catalyzed by uv light or base, and are thought to occur via S—S bond fission, since the radiotracer experiments of Guryanova *et al.* [4, 5] show that no C—S bonds are broken in EtS_2*SEt exchange with Et_2S_4. The trisulfide exchange reaction given in Eq. (9.1) has an activation energy of 122 kJ mole^{-1} as measured in the temperature range 132–148°, and is subject to autocatalysis from the tetrasulfide product of the subsequent reaction in Eq. (9.2) which also occurs in the system [6]. Simplified kinetics are obtained if 0.3 % of the tetrasulfide is added initially as catalyst. A free radical chain mechanism would account for the observed activation energy, which is about one half of the expected bond-dissociation energy for the S—S bond broken in Eq. (9.3) and free radical carriers [7] are common in reactions of sulfur chains. As expected for such a mechanism, rates are virtually unaffected by change in solvent polarity.

$$Et_2S_3 \rightleftharpoons EtS\cdot + \cdot SSEt, \quad \text{initiation} \qquad 9.3$$

$$EtSS\cdot + n\text{-PrSS-}n\text{-Pr} \rightarrow EtSSS\text{-}n\text{-Pr} + n\text{-PrS}\cdot, \quad \text{propagation} \qquad 9.4$$

The rate of reaction in Eq. 9.1 is much higher than that of the disproportionation, Eq. 9.2, so it is clear that the fastest chain step would involve attack of the chain carrier on the central sulfur of an unreacted substrate molecule. The central sulfur would be most prone to homolytic attack if the combined substituent effects of the alkyl terminal groups and the adjacent sulfur atoms are considered. The complete mechanism proposed [6] has three possible initiation steps arising for homolysis of Et_2S_3,

n-Pr$_2$S$_3$, and Et-n-PrS$_3$, then six possible carrying steps, and three possible termination steps. From this, using the steady state approximation, an expression for the overall rate is developed with which the experimental rate data are consistent.

The second type of exchange, Eq. 9.2, involves exchange of chain fragments of different lengths to produce polysulfides of random chain length. Grant and Van Wazer [8] produced such systems by equilibrating dialkyl disulfides and sulfur S$_8$ molecules at various temperatures and studied the inherent reactions by means of the ^{1}H NMR signals of the alkyl groups. The time required for equilibration depended on the nature of the alkyl group used and on the ratio of alkyl sulfide to sulfur molecules in the initial composition of the mixture. Sulfur-rich mixtures and bulky alkyl groups slowed down the reaction considerably, with 40 days at 118° being required for the composition [methyl]/[sulfur] = 0.2. Sulfur is incorporated into the dialkyl sulfide chain to give mixtures of polysulfides of various chain lengths. Those sulfides of chain length up to six sulfur atoms (occasionally up to ten atoms) long could be distinguished and their concentration determined via high resolution ^{1}H NMR signals of the alkyl groups. With this data the equilibrium between different sized chains was quantitatively studied at 118°. The mechanism of the reaction (S$_8$ + R$_2$S$_2$) is considered to require a free radical in view of the normal reactions of sulfur molecules at these temperatures [7], involving reaction sequences such as Eqs. 9.5–9.7.

$$RSSR \rightleftharpoons 2RS\cdot, \quad \text{initiation} \qquad\qquad 9.5$$

$$RS\cdot + S_8 \rightarrow RS_9\cdot, \quad \text{propagation} \qquad\qquad 9.6$$

$$RS\cdot + RS_9\cdot \rightarrow RS_{10}R, \quad \text{termination, etc.} \qquad\qquad 9.7$$

The progress of the reaction was monitored by ^{1}H NMR and dialkyl disulfide was seen to decay rapidly due to a buildup of chains with six or more sulfur atoms. Subsequently, redistribution reactions between chains occurred to give tri-, tetra-, and pentasulfides, but these are not formed randomly. The dimethyl trisulfides are present in greater and the disulfides in less than the statistical amount. A representative set of equilibrium compositions is given in Table 9.2.

Pickering *et al.* [9] examined some reactions of type 9.2 directly, using the NMR techniques developed by Grant and Van Wazer [8]. Dimethyl tetrasulfide decomposed rapidly at 80° in a radical chain reaction, set off by a symmetrical homolytic fission [8]. The rate of this step is already known [10]. The addition of Banfield's free radical inhibited the initial decay of Me$_2$S$_4$ until all the radical had been used up. Unsymmetrical cleavage of the

TABLE 9.2

EQUILIBRIUM OF $RS_2R + S_8{}^a$

Mole ratio R/S	R	Values of n				
		2	3	4	5	6
0.89	Me	75	22	3	0	0
0.87	t-Bu	77	14	5	2	1
0.52	Me	21	29	22	13	15
0.52	t-Bu	49	10	15	8	18
0.47	Me	14	24	21	15	26
0.48	t-Bu	43	9	15	8	28
0.89	Statistical	80.2	15.9	3.2	0.6	0.1
0.52	Statistical	35.1	22.8	14.8	9.6	17.7
0.48	Statistical	31.6	21.6	14.8	10.1	20.9

[a] Mole percentage of RS_nR in product at 120°.[b]
[b] From Grant and Van Wazer [8].

tetrasulfide can be excluded as an initiating step since no dimethyl disulfide is observed in the early stages of the reaction. The decomposition of dimethyl trisulfide under the same conditions is much slower, but does give rise to dimethyl disulfide. It is possible that $MeS_x \cdot$ radicals are stabilized when $x = 2$ by the extra sulfur atoms in the chain. It is known that the S—S bond energy [10] is 151 kJ mole^{-1} in Me_2S_4 and 139 kJ mole^{-1} in polymeric sulfur, but about 294 kJ mole^{-1} in disulfides. The unsymmetrical cleavage of Me_2S_4 would thus break a stronger bond, and is less likely to occur than 9.8, the symmetrical cleavage.

$$Me_2S_4 \rightarrow 2MeS_2 \cdot \qquad 9.8$$

The final composition of the tetrasulfide reaction mixture after several half-lives was 46% trisulfide, 24% tetra-, 13% penta-, 11% hexa-, and 6% disulfide. This may not be completely equilibrated, since agreement with the appropriate data in Table 9.2 is not good. Propagation steps such as Eqs. 9.9–9.11 are proposed to account for the products.

$$MeSSSSMe + MeS_2 \cdot \rightarrow MeSSSMe + MeS_3 \cdot \qquad 9.9$$

$$MeSSSSMe + MeS_3 \cdot \rightarrow MeS_5Me + MeS_2 \cdot \qquad 9.10$$

$$2MeS_3 \cdot \rightleftharpoons Me_2S_6 \qquad 9.11$$

Isomerization through the intermediacy of a branched sulfur chain has

$$-S-S- \rightleftharpoons -S=S$$

also been observed [11] in di- and polysulfide chains. This occurred much faster than reactions 9.1 or 9.2 and was detected through a stereochemical label. Compound I, bis(1,3-dimethylbut-2-enyl) trisulfide, exists in meso and racemic forms. A single isomer of compound I was isolated and observed to isomerize in a first-order process to an equal mixture of racemic and meso forms, with a half-life of 46.5 min at 75°. Since no mixed trisulfides were obtained (reaction 9.1) in the presence of other R_2S_3 entities, nor any disproportionation products (reaction 9.2) it was proposed that racemization of one optically active carbon occurs through the branched sulfur chain intermediate II which rapidly reverts to the mixture of isomers of compound I in equal amounts.

I   II*

* The curving arrow in II denotes free rotation.

Mixtures of dimethyl disulfide and dichlorodisulfides provide a scrambling reaction after an initial evolution of alkyl chloride. Long sulfur chains are built up with terminal alkyl or halo groups, and these equilibrate randomly after the fashion of Eq. 9.2 to give chains of various sizes [8].

The simplest reaction in this series does not involve disproportionation at all. This is the exchange of hydrogen and deuterium on sulfur (Eq. 9.12).

$$H_2S + D_2S \rightleftharpoons 2HDS \qquad\qquad 9.12$$

It is thought to be random and some limits to its rate can be fixed from the ^{1}H NMR study of Takahashi and Hazato, who found an exchange rate slower than 0.5 sec^{-1} between HDS and H_2S at 24° [12].

Mixtures of sulfanes H_2S_x are produced on hydrolysis of alkali polysulfides with acid [13, 14]. These undergo "cracking" on distillation and chains of more than six sulfur atoms cannot be distilled for this reason.

Chains with $x \geqslant 7$ are obtained from the reaction of chlorosulfanes and sulfanes [15–17] shown in Eq. 9.13. The rate of reaction decreases in the

$$m\text{H}_2\text{S}_x + n\text{Cl}_2\text{S}_y \rightleftharpoons (m - n)\text{H}_2\text{S}_z + 2n\text{HCl} \qquad 9.13$$

order $x = 2 > 3 > 4$, etc., and the desired product is obtained by removal of HCl. The asymmetric ClS_nH are not formed, although RS_nH and RS_nCl do exist under other conditions [18].

B. Sulfur(IV)

Sulfur dioxide is found to undergo oxygen exchange at room temperature [19], both in the gas phase and in moist carbon tetrachloride solution. The exchange of samples containing different oxygen isotopes (Eq. 9.14) was

$$\text{S}^{18}\text{O}_2 + \text{S}^{16}\text{O}_2 \rightleftharpoons 2\text{S}^{18}\text{O}^{16}\text{O} \qquad 9.14$$

complete in a few minutes and the equilibrium constant for Eq. 9.14 was about 3. The mechanism is not certain; surface catalysis by the Pyrex reaction vessel, or catalysis by traces of moisture were possible. In the absence of these it is unlikely that SO_2 could react by an ionic process, so a four-center mechanism is possible [19]. The mixed halides thionyl chloride fluoride [20] and thionyl bromide fluoride [21] are known, but nothing is known of the equilibrium with pure halides. From mixtures of thionyl chloride with thionyl bromide, no mixed halide could be isolated on distillation [22] but it was found that a sulfur isotopic label ^{35}S was readily transferred at temperatures of $-50°\text{C}$ between chloride and bromide, Eq. (9.15), and at

$$^*\text{SOCl}_2 + \text{SOBr}_2 \rightleftharpoons \text{SOCl}_2 + {^*\text{SOBr}_2} \qquad 9.15$$

$-80°$ the reaction was slow enough for its rate to be measured. This sulfur exchange is surely evidence of a mixed halide intermediate, and of a rapid scrambling reaction, such that distillation results in the separation of the least (bromide) and most (chloride) volatile thionyl halides. Probably the mixed halide could be detected by vibrational spectroscopy.

C. Sulfur(IV) and Sulfur(VI)

Thionyl and sulfuryl chlorides exchanged sulfur (^{35}S) extremely slowly (no exchange in 5 days at $0°$). Exchange of halogen in this system could not, of course, result in effective sulfur exchange, so the ionic and four-center mechanisms are rejected. The authors suggest a rate-determining formation of a symmetrical adduct $\text{Cl}_2\text{OS}—\text{O}—\text{SOCl}_2$ in which the sulfuryl chloride has to act as oxide donor (base) to thionyl chloride [22].

D. Sulfur(VI)

Sulfur trioxide reacts with dimethylsulfate within a day at 24°C, yielding an equilibrium mixture of SO_3 and SO_3 chains terminated by methoxy groups (polysulfate esters) as shown in Eq. 9.16. Equilibrium constants for

$$SO_3 + (MeO)_2SO_2 \rightleftharpoons MeO(SO_3)_nSO_2Me \qquad 9.16$$

the general redistribution reaction, Eq. 9.17, taking place in the mixture should statistically be unity, but it appears that all end and middle SO_3 units are not equally probable [23].

$$2MeO(SO_3)_mMe \rightleftharpoons MeO(SO_3)_{m-1}Me + MeO(SO_3)_{m+1}Me \qquad 9.17$$

The equilibria can conveniently be investigated *in situ* using the technique of ^1H NMR spectroscopy and the equilibrium constants K obtained for reaction 9.17 differ greatly from the random value unity, where $m = 2$, 3, or 4, as will be seen by inspection of Table 9.3. For larger chains the random

TABLE 9.3

EQUILIBRIUM CONSTANTS, K, FOR REACTION 9.17 AT $72°$ [a]

m	K	ΔF (kJ mole^{-1})	m	K
2	0.02 ± 0.009 [b]	5.44	5	0.91 ± 0.18
3	0.153 ± 0.04	2.52	6	1.1 ± 0.2
4	0.70 ± 0.012	0.42	7	0.8 ± 0.2

[a] From Van Wazer *et al.* [23].
[b] Standard deviation of at least six runs.

value applies within reasonable limits. A reorganization heat order of 4 is necessary to explain the deviations from unity [24]. This means simply that substituents up to four SO_3 units distant influence the enthalpy of reaction (9.17). The equilibrium between SO_3 and the chains is also affected by the chain substituents [23]. Equation 9.18 applies at $72°$ (monomer SO_3).

$$MeO(SO_3)_mMe \rightleftharpoons MeO(SO_3)_{m-1}Me + SO_3 \qquad 9.18$$

When $m = 6$, the equilibrium constants K_m for Eq. 9.18 should be independent of the value of m. Experimental values of K_m given in Table 9.4 support this view [23]. Data at 24° take into account also the equilibrium between monomer and trimer SO_3, which renders Eq. 9.19 important at this temperature, and Table 9.4 contains data at 24° for Eqs. 9.18 and 9.19. A

$$MeO(SO_3)_m Me \rightleftharpoons MeO(SO_3)_{m-3} Me + (SO_3)_3 \qquad 9.19$$

reorganization enthalpy for Eq. 9.18 of approximately 8.4 kJ mole^{-1} can be obtained from this table.

Polysulfuryl fluorides chains redistribute nonrandomly among themselves and the strong effect of chain substituents several atoms removed from the site of exchange is again responsible [23].

TABLE 9.4

VALUES[a] OF K_m IN mole liter^{-1} FOR EQS. 9.18 AND 9.19

	K_6	K_7	K_8	K_9	Temp (°C)
Eq. 9.18	0.025	0.021	0.023	0.029	72
Eq. 9.18	0.017	0.013	0.015	0.014	24
Eq. 9.19	0.045	0.021	0.028	0.029	24

[a] From these values ΔH for Eq. 9.18 is ~8.4 kJ mole^{-1}.

Dean and Gillespie [23a] have recently observed the formation of poly-fluorosulfuric acids $H(SO_3)_n F$ in solutions of sulfur trioxide (SO_3) in fluorosulfuric acid (HSO_3F) or fluorosulfuric acid containing the diluent SO_2ClF. At temperatures below $-100°$ this diluent gives less viscous mixtures than neat fluorosulfuric acid. At $-110°$ the ^{19}F NMR spectrum of a solution containing $SO_3 : HSO_3F : SO_2ClF$ in the mole ratio 1.94 : 1.00 : 2.85 shows a single line from the diluent and seven other lines for the various $-SO_3F$ in the polyfluorosulfuric acids. From the peak areas of ^{19}F signals in this and other mixtures, the equilibrium constants K_n for the reaction 9.20, where $n = 2$–4, were found to be $K_2 = 0.14$–0.16, $K_3 = 0.55$–0.54, and $K_4 = 0.8$, approaching the value of unity as n increases. Compare the data on methyl polysulfates [23].

$$2H(SO_3)_n F \rightleftharpoons H(SO_3)_{n-1} F + H(SO_3)_{n+1} F \qquad 9.20$$

II. Selenium, Tellurium, Polonium, and Mixtures

Scrambling of alkyl and halogen groups on polyselenide chains is one of several reactions which can take place on mixing dimethyldiselenide and dichlorodiselenide [25]. The first reaction is elimination of alkyl chloride to give alkyl- and chloro-terminated polyselenides, then scrambling of end groups, then scrambling of chains to give a distribution of chain lengths occur. Selenium also precipitates. The situation is similar to that for reactions of alkyl sulfides and sulfur dichloride already described (Section I). Initially dialkyl chains build up and then rapidly decay in favor of mixed alkyl halogen chains [25].

$$R_2Se_n + Se_2Cl_2 \rightleftharpoons RSe_nCl + RSeCl, \quad etc.$$

The mixed sulfide–selenide $(C_6F_5)_2SSe$ has been made by the direct redistribution of $(C_6F_5)_2S_2$ and $(C_6F_5)_2Se_2$ and is thermodynamically favored [26]. Mixed sulfoselenides and -tellurides are also known [27].

The mixed halides of tellurium ($TeCl_2I_2$ and $TeBr_2I_2$) and the corresponding halides of polonium have been reported [28, 29] in the solid state but nothing is known of their possible disproportionation to homohalides.

REFERENCES

1. D. T. McAllan, T. V. Cullum, R. A. Dean, and F. A. Fidler, *J. Amer. Chem. Soc.* **73**, 3627 (1951).
2. S. F. Birch, T. V. Cullum, and R. A. Dean, *J. Inst. Petrol.* **39**, 206 (1953).
3. L. Haraldson, C. J. Olander, S. Sunner, and E. Varde, *Acta Chem. Scand.* **14**, 1509 (1960).
4. E. N. Guryanova, Y. K. Sirkin, and L. S. Kuzina, *Dokl. Akad. Nauk. SSSR* **86**, 107 (1952); *Chem. Abstr.* **47**, 1457f (1953).
5. E. N. Guryanova, L. A. Egovora, *Zh. Obsch. Khim.* **28**, 1745 (1958); *Chem. Abstr.* **53**, 1108d (1959).
6. C. D. Trivette, Jr., and A. Y. Coran, *J. Org. Chem.* **31**, 100 (1966).
7. G. K. Fraenkel and D. M. Gardner, *J. Amer. Chem. Soc.* **78**, 3279 (1956).
8. D. Grant and J. R. Van Wazer, *J. Amer. Chem. Soc.* **86**, 3012 (1964).
9. T. L. Pickering, K. J. Saunders, and A. V. Tobolsky, *J. Amer. Chem. Soc.* **89**, 2364 (1967).
10. I. Kende, T. L. Pickering, and A. V. Tobolsky, *J. Amer. Chem. Soc.* **87**, 5582 (1965).
11. D. Barnard, T. H. Houseman, M. Porter, and B. K. Tidd, *Chem. Commun.* **1969**, 371.
12. K. Takahashi and G. Hazato, *Bull. Chem. Soc. Japan* **38**, 1807 (1965).
13. G. Schwarzenbach and A. Fischer, *Helv. Chim. Acta* **43**, 1365 (1960).
14. F. Fehér, W. Laue and G. Winkhaus, *Z. Anorg. Allg. Chem.* **288**, 113 (1956).
15. F. Fehér and S. Ristic, *Z. Anorg. Allg. Chem.* **293**, 307 (1957).
16. F. Fehér and G. Winkhaus, *Z. Anorg. Allg. Chem.* **288**, 123 (1956).

17. F. Fehér and W. Laue, *Z. Anorg. Allg. Chem.* **287**, 45 (1956).
18. G. Nickless, "Inorganic Sulphur Chemistry." Elsevier, Amsterdam, 1968.
19. N. N. Lichtin, J. Laulicht, and S. Pinchas, *Inorg. Chem.* **3**, 537 (1964).
20. R. J. Gillespie and E. A. Robinson, *Can. J. Chem.* **39**, 2179 (1961).
21. H. Jonas, *Z. Anorg. Allg. Chem.* **265**, 273 (1951).
22. L. Johnson and T. H. Norris, *J. Amer. Chem. Soc.* **79**, 1584 (1957).
23. J. R. Van Wazer, D. Grant, and C. H. Dungan, *J. Amer. Chem. Soc.* **87**, 3333 (1965).
23a. P. A. W. Dean and R. J. Gillespie, *J. Amer. Chem. Soc.* **92**, 2362 (1970).
24. D. W. Matula, L. C. D. Groenweghe, and J. R. Van Wazer, *J. Chem. Phys.* **41**, 3105 (1964).
25. D. Grant and J. R. Van Wazer, *J. Amer. Chem. Soc.* **86**, 3012 (1964).
26. E. Kostiner, M. L. N. Reddy, D. S. Urch, and A. G. Massey, *J. Organometal. Chem.* **15**, 383 (1968).
27. A. J. Parker, *Acta Chem. Scand.* **16**, 855 (1962).
28. E. E. Aynsley, *J. Chem. Soc.* **1953**, 3016.
29. K. W. Bagnall, R. W. M. D'Eye, and J. H. Freeman, *J. Chem. Soc.* **1956**, 3385.

10

Trends in Redistribution Reactions

The accumulation of kinetic and thermodynamic data has been particularly extensive in main group chemistry over the last decade. There are signs of a growing awareness of the occurrence of redistribution phenomena in transition metal chemistry and many new developments are to be expected in this area, particularly when one bears in mind the variety of ligands which will be available for study in this region of the periodic table.

The thermodynamics of scrambling of monofunctional substituents have been found to be largely dependent on the scrambled group rather than on the central atom. This observation extends now to B≡, PhB≡, i-BuB≡, Si≡, MeSi≡, Me₂Si≡, Ge≡, MeGe≡, Me₂Ge≡, Sn≡, P≡, OP≡, SP≡, As≡, R₃As≡, Sb≡, R₃Sb≡, S≡, and Se≡ among main group elements, to Ni(II) in transition metal chemistry, and to scrambling of pairs of substituents such as Me–Et, OEt–OMe, tfac–hfac, MeO–Cl, NMe₂–Cl, PF₃–CO. All the substituents mentioned have not, of course, been studied for every central atom mentioned. The general information available now is sufficiently great that prediction of the free energy of many reactions can be made qualitatively. A caveat is necessary here. Scrambling of two ligands may normally be random, but in the presence of differential π bonding become strongly endo- or exothermic. This is the case, for instance, for scrambling of halogen (normally random) when distributed between the

157

two apparently similar central groups Me_2Si= and Me_3Si— (Chapter 7). It is also the case for PF_3 and CO, random under normal circumstances, but not when the central group is five-coordinate HCo≡ (PF_3 is strongly favored due to preferential bonding in certain stereoisomers; see Chapter 5).

Kinetics are much more dependent on the central metal atom than thermodynamics. In general, the better the central atom can confer Lewis acidity on a compound, the more rapid its scrambling reactions are likely to be. Among exchanging substituents, donor power is a gauge of kinetic lability. Reactions are usually bimolecular or consist of more than one bimolecular sequence, each of which creates two bonds, and breaks two bonds. Mechanisms are frequently discussed in terms of the "four-center" path (I) which can be adapted to permit any sequence of steps, a, b, c, d and

$$MX_n + MY_n \rightleftharpoons (X)_{n-1}M \overset{d \diagdown \ \ X \diagup a}{\underset{c \diagup \ \ Y \diagdown b}{\diagdown \diagup}} M(Y)_{n-1} \rightleftharpoons MX_{n-1}Y + MXY_{n-1}$$

I

any method of bond fission (free radical, heterolytic, or molecular dissociation). Free radical fission has definitely been established in sulfur chemistry (Chapter 9), heterolytic fission in mercury reactions (Chapter 4, Section V), and molecular dissociation is probable in some lithium "ate" complex exchanges (Chapter 3) and in transition metal carbonyls (Chapter 5). Concerted reactions (a, b, c, d virtually synchronous) seem to be prevalent, and have usually been marked by large negative entropies of activation, as in titanium chemistry (Chapter 5, Section IV), antimony chemistry (Chapter 8, Section III). In some of the slower reactions there is evidence that hydrolysis or catalytic amounts of impurity accelerate reaction—presumably because the catalytic path is faster than any four-center one. This is especially true of the very slow reactions which occur in some phosphate and silicon series, but it is also true of systems which have extreme sensitivity to moisture, like the halogen exchange between $GeCl_4$ and BBr_3. Other catalysts of importance are the very strong Lewis acids of Group III, e.g., aluminum bromide and chloride. These presumably are catalytic by virtue of this Lewis acidity which suggests a four-center path.

In the early main groups (I, II, III) where acceptor tendencies are high, four-center mechanisms are reasonable, and are definitely established for some reactions. Where bridging dimers exist, an insight into mechanism has been possible (aluminum and boron; see Chapter 6).

An area of transition metal chemistry which should be open to mechanistic study is the exchange of π-bonding ligands such as CO, PF_3 on structures of high coordination numbers, where the existence of stereoisomers may permit attack on the steric course of the reaction to be launched. There is also a tremendous number of ligands in which structural variations are possible, and which will enable linear free energy correlations to be investigated for transition metal redistribution reactions.

The studies in polymeric systems which have been made are mostly in main group chemistry and are the work of Van Wazer and his associates. A particularly important outcome of this work is the general theory of structural reorganization, which enables a mathematical description of the thermodynamics of polymer formation [1] in terms of chain, branched and cyclic species. It is now feasible to make predictions from intersystem scrambling of monofunctional substituents, of the general composition to be expected in the related polymer families where one monofunctional substituent is replaced by a bridging one [2]. This has tremendous implications for inorganic polymer chemistry and must stimulate a great deal of new work. The introduction of a terminal group X in a system of two central atoms and bridging Y, where X is not compatible with random Y–X exchange can push the whole system over to a specific polymer form.

Spinoff

The great upsurge in production of data on scrambling reactions has enabled correlations of various physical properties (such as NMR chemical shift or vibrational frequency data) with calculated parameters for mixed compounds. This should continue to be a most profitable area of research. Relations between free energies and substituents should also be considerably investigated in future, to provide qualitative structure–reactivity correlations like the Hammett $\sigma\rho$ relation for aromatic reactivity. A start has been made by Van Wazer and Moedritzer but much more data is needed. Another area where experimental data and theoretical computations thereof can be compared to mutual advantage is the calculation of free energies for scrambling reactions *a priori*. Some effort in this area has been made [3].

Correlation of observed vibrational frequencies in simple compounds such as mixed boron halides [4] or hydride halides [5] with frequencies calculated using simple force field calculations or the Green's function method have been made.

The NMR chemical shifts of the central atom in a redistribution reaction are affected by substituents on all sides and so are sensitive functions of the substituents. In the literature is a large body of chemical shift data for mixed compounds which could be used to test theoretical interpretations of chemical shifts. Van Wazer and Letcher have recently made extensive correlations of ^{31}P chemical shift data with some calculated parameters. There are considerable deviations from additivity which can be calculated by summing a contribution to the shift from both σ and π bonding of substituents to central phosphorus. There is a generally negative deviation from linearity in shifts for mixed compounds in a series, e.g., MPX_3, MPX_2Y, $MPXY_2$, and MPY_3. The deviation is appreciably dependent on the π character of bonding in the mixed compounds [6]. Central atoms whose shifts might be studied include ^7Li, ^{11}B, ^{15}N, ^{27}Al, ^{31}P, ^{119}Sn, ^{55}Mn, and ^{131}Tl. Shifts of ^1H and ^{19}F resonances are affected in a one-sided manner by scrambling reactions. Where these occur in scrambling reactions they occur only in attached groups, so they are much less sensitive to variations in other substituents in mixed compounds.

REFERENCES

1. D. W. Matula, L. C. D. Groenweghe, and J. R. Van Wazer, *J. Chem. Phys.* **41**, 3105 (1964).
2. J. R. Van Wazer and K. Moedritzer, *J. Amer. Chem. Soc.* **90**, 47 (1968).
3. I. Eliezer and Y. Marcus, *Coord. Chem. Rev.* **4**, 273 (1969).
4. L. P. Lindemann and M. K. Wilson, *J. Chem. Phys.* **24**, 242 (1956).
5. T. Wolfram and R. E. DeWames, *Bull. Chem. Soc. Japan* **39**, 207 (1966).
6. J. R. Van Wazer and J. H. Letcher, *in* "Topics in Phosphorus Chemistry," Vol. 5, Chapter 3. Wiley (Interscience), New York, 1967.

Author Index

Numbers in parentheses are reference numbers and indicate that an author's work is referred to although his name is not cited in the text. Numbers in italics show the page on which the complete reference is listed.

Subject Index

A

Acetylacetonate, *see also* β-Diketonates
scrambling on Zr, 2

Acid clays as scrambling catalysts for siloxanes, 116

Alkyl exchange, *see also* specific elements
between Cd and Ga, In, Zn methyls, 60

Aluminum, 93–101
bridged dimer as transition state, 93
exchange of acetylacetonate and DMF, 99
intermolecular exchange in trimethyl-alane, 93
mechanisms for exchange of groups between dimers, 98
mixed dimers of, 96, 99
mixed tetrahaloaluminates and ^{27}Al NMR, 99

Aminotroponiminates, 68
exchange on nickel, 68

Antimony, 143–145
exchange of halogen in organoantimony dihalides, 144
kinetics of methyl–chloride exchange, 145
mixed halide, 143

Arsenic, 138–143
cage compounds as sources of bridging atoms, 141
exchange of halogen with alkoxide, dialkyl-amino or phenyl, 139
with organoarsenic dihalide, 140, 141
mixed halides of, 138, 140, 141
reorganization with bridging nitrogen, 141
with bridging oxygen, 141–143

Ate complexes, 37–39
of Hg and Li, 38, 39
of Mg and Li, 37, 38
of Zn and Li, 37

B

Beryllium, 44–45
exchange of DMF and acetylacetonate on, 44
Schlenk equilibrium in, 45

Bis(1,3-dimethylbut-2-enyl)trisulfide, iso-merization by first-order process, 151

Bismuth, 145
mixed trihalides, 145

Borazoles, scrambling of substituents, 92, 93

169